Helen Adrienne LCSW

ISBN: 1452853754
ISBN-13: 9781452853758
Library of Congress Control Number: 2010906585

# DEDICATION

To all the babies who might not have been born…and their parents.

# Table of Contents

# List of Illustrations

Helen Adrienne is an award-winning, licensed psychotherapist with more than thirty years of experience in the field of infertility. She trained in mind/body therapy at Harvard's Benson-Henry Mind/Body Medical Institute and trained in clinical hypnotherapy at the New York Society for Ericksonian Psychotherapy and Hypnosis (NYSEPH). She is a certified hypnotherapist and approved consultant for the American Society of Clinical Hypnosis. Besides being a clinician, presenter on infertility to national and international professional and lay audiences, and author of many articles on infertility, Adrienne teaches hypnosis for NYSEPH, and organizes master classes in clinical hypnotherapy taught by Jeffrey Zeig, PhD, founder of the Milton H. Erickson Foundation. She also conducts mind/body stress reduction groups for the wellness program at the prestigious NYU Fertility Center in New York City. Ms. Adrienne, recognized as an expert in the field of infertility, has been featured on various TV and radio shows and has been interviewed for print and online magazines. She lives and works in New York City.

# Endorsements

"*On Fertile Ground* is a much-needed book in a field where the deep psychological processes of the infertile patient are often forgotten. Based on over thirty years of experience in the field of psychotherapy, the author, Helen Adrienne, offers wise guidance and skilled hypnotherapy to women and men who thought that giving birth to their own child was an impossible dream. Read and believe."

— *Chris Page, MD, international speaker and author of* Frontiers of Health: How to Heal the Whole Person *and* Spiritual Alchemy.

"Helen Adrienne is a talented teacher and clinician, and *On Fertile Ground* is a groundbreaking book that can help the millions who are trying to conceive. Adrienne is perceptive, brilliant, and compassionate. She offers a new perspective on helping infertile couples—including sure-fire methods that guide these troubled couples through the maze of problems that they confront."

—*Jeffrey K. Zeig, PhD, founder and director, The Milton Erickson Foundation.*

"Reading this book is an investment not only in coping with infertility, but it also provides the awareness and strength to confidently deal with the daily confrontations life presents. With a focus on infertility, the book will bring clarity to the complicated demands of work, relationships, and self. *On Fertile Ground* provides readers with an antidote for the destructive forces related to infertility and pregnancy loss."

—*Dr. Frederick Licciardi, reproductive endocrinologist and head of the Wellness Program at the NYU Fertility Center, New York, NY.*

"Helen Adrienne is on the cutting edge of adjunctive mind-body infertility treatment. A gifted hypnotherapist, Adrienne's new book, *On Fertile Ground,* gives a compassionate understanding of a complex problem and is a must read for anyone struggling with this painful issue and for those who care about them."

—*Suzanne Drake, PhD, APN-C family therapist and childbirth educator.*

"Helen Adrienne is the finest clinician I know in the field of infertility. She is wise, compassionate, and creative, and her deep understanding of the issues facing the infertile couple is expressed in her new book, *On Fertile Ground.* She has a wide range of impressive skills that she has carefully developed and honed to work with this population."

—*Susan Dowell, LCSW*

Helen Adrienne, LCSW , BCD, a licensed clinician and certified hypnotherapist, has several decades of experience working with infertile women using hypnotic techniques, and in *On Fertile Ground* she describes how one can incorporate self-hypnosis safely and effectively in a condition that affects millions of couples worldwide. As a practicing reproductive endocrinologist and infertility specialist, and director of a mind-body medicine program for couples with infertility issues, I am very delighted that women who are having trouble conceiving now have this easy-to-access guide to help alleviate their stress and possibly improve the likelihood of achieving their goal."

—*Brinda Kalro, MD, reproductive endocrinologist, Magee Women's Hospital, Pittsburgh, PA.*

"Helen Adrienne's book, *On Fertile Ground: Healing Infertility,* is bound to be groundbreaking. Until now, most of the solutions to infertility have been medically oriented, draining couples of their financial, physical, and emotional resources. This timely book sheds a fresh new light on infertility in all its complexities. Based on a mind/body approach, it plumbs the depths of the beliefs and stigma of being infertile, the effect on intimate relationships, and the hopes, dreams, anguish, despair, and joy that accompany it. Adrienne provides new and practical guidelines for dealing with all of these aspects. The stories she tells of her own and her patient's experiences are fascinating and heart rending. The book should be helpful, not only to millions of couples around the world, but to the helping professions as well as the lay public."

—*Peggy Papp, MSW, director of family therapy, Ackerman Family Institute and Mt. Sinai Adolescent Clinic; author of* The Process of Change, *co-author of* The Invisible Web: Gender Patterns in Families; *and editor of* Couples on the Fault Line: New Directions for Therapists.

"It has been my honor to know Helen as an experienced and talented clinician. *On Fertile Ground* is a book that is sensitive to the needs of anyone coping with the emotional distress of infertility. The fact that she is incorporating the application of self-hypnosis as a healing tool will make it a particularly unique and powerful resource."

—*Carolyn Daitch, PhD, director, Center for the Treatment of Anxiety Disorders; author of* Affect Regulation Toolbox.

"Helen Adrienne's compassionate and caring work is unique and unparalleled in the field of infertility. Her approach to her patients builds on years of work as a therapist and a continuing desire to do more, and better, for those who seek her guidance. In *On Fertile Ground,* Adrienne pushes the envelope, combining traditional and nontraditional therapies to achieve the desired results for her patients."

—*Abigail Brenner, MD, psychiatrist and author of* Transitions: How Women Embrace Change and Celebrate Life.

"My patients who have seen Helen Adrienne have all come out of counseling more confident and easier to care for."

*—Richard Lumiere, MD, ob-gyn, in private practice, New York City.*

# Acknowledgments

Life is about fertility, both literal and figurative. For three decades, this book lay in a fallow field within me, all the while gathering nutrients. These acknowledgments are to thank the people who enriched my soil, so that this book could take root and grow.

I shall be eternally grateful to Christine Page, MD, whose confidence in me provided the most potent fertilizer and burst open the seed of *On Fertile Ground.* Thanks, also, to her husband, Leland Landry, for his hardy and hearty "Amen."

I am blessed to have many friends and colleagues who tilled the soil. Suzanne Drake, my dearest friend and colleague, has always supported my aspirations—this book being no exception. Abigail Brenner provided the title, which says everything that I hoped to express in three words. She, Carolyn Daitch, and Peggy Papp, all of whom have written books before me, were beacons of sunshine whose warmth was a constant reminder that I could grow through the disorientation of the creative process. Susan Dowell was an every-ready sounding board. And, a special thanks goes to Jeff Zeig, who saw the flower in me when I was only a bud.

From Lois Nachame, I learned that writing is a team sport. I was thus inspired to ask various people to evaluate drafts of my work along the way. Karen Beck, Carol Ciacci, Julie Hall, Peggy Papp, and Susan Dowell generously gave their time. I am particularly grateful to Monique, Julie, Jenn, Andrea, Stephanie, Monica, Elizabeth, and Cindy, whose suggestions were invaluable because they represent my target audience.

Thanks to Marilyn Fish, whose wit and intelligence has always figured in my projects, and Ann Webster, PhD, whose capacity for joy could buoy up a lead balloon. So many other women friends, new and old, have encircled me with the kind of love and support that translates to compost: Annie Cunningham, Paula Esposito, Rita Fertel, Carol Fitzsimons, Teresa Garcia, Ellen Jacobson, Maddy Katz, Vivian Leyton, Florence Manoff, Annie Pearl, Christine Ranck, Phyllis Roberts, Margie Rosenberg, Rita Sherr, Marcia Siegel, and Vivian Yaker. I owe Mitchell Fink and Rick Miller gratitude for supplying valuable male energy. Thanks to my kindred spirits at Resolve: Janet

Lista, Phyllis Lowinger, Anne Malave, Ellen Shapiro, Elizabeth Silk, and Mary Fraoli. Special thanks to Resolve officers Sue Slotnick and Dawn Gannon for their personal openness and availability to me.

In case I was at risk to forget that our challenges teach us how to grow and mature, thanks go to my sister, Claire and her family, who continue to display a dignity and grace in the face of adversity that would inspire anyone dodging life's bullets.

The preciousness of having my own daughters and sons-in-laws, Kim, Deb, Rob, Glenn, and Brett, and grandchildren, Jaden, Macy, Reese, and Brooke, make it easy for me to understand the longing of those whose goal to have a family has been thwarted, temporarily or otherwise.

Thanks to my extended family members, the Kantrowitz and Harris clans, whose interest in my work has been like a quenching rain. The same goes for the Cohns, Rosenbergs, Spurrs, Novicks, Sarasohns, Rusoffs and Lancins.

To my Master Class "Family"—there could never be a better setting for cross-pollination. Thank you for the atmosphere of "Yes, you can!"

Many thanks to two medical colleagues at NYU Fertility Center: Fred Licciardi, MD, who fact checked the language of reproductive endocrinology for its suitability for the lay public and who, by reading the book in its entirety, confirmed that it would be of interest to the medical community as well. The time you invested is greatly appreciated. Thanks, also, to Jamie Grifo, MD, medical director at the center. You made the exposure of my work to a larger audience possible, which means the world to me.

This book would have fallen flat without the participation of all of my interviewees. Your willingness to share your stories gave life to the message of this book—that there is fertility to be found in infertility. I am grateful to the thousands of infertility patients with whom I've had the privilege to work over the years. You've taught me everything that I needed to know in order to serve your needs.

And last, but far from least, thanks to Marc Spurr, who taught me that dreams could come true.

# Preface

This book springs from a well deep within me. It is *my* well, but it is fed by a universal well, originating from the pure waters of femininity that are a comingling of womanhood and motherhood. I wrote it from a tide of compassion for the infertile patient, with whom I have had the privilege of working for over three decades as a psychotherapist.

You might wonder why I would call a book meant for those dealing with the nightmare of infertility *On Fertile Ground.* It is because I have repeatedly experienced in my own growth and development, that our emotional nightmares are our teachers. If we are willing to look this nightmare in the eye, it is fertile with possibilities for growth.

If we let them, nightmares can free us to discover our higher potential. They can crack us open in the best sense. There is a tree that grows in Yellowstone National Park, the lodge-pole pine tree, that needs the intense heat of a forest fire to crack open its seed, so that new life can take root in the devastated but cleared space. The inferno prepares the soil and the seed for each other.

Nightmares are the psyche's way of communicating with us, albeit in symbols. Taken at face value nightmares are frightening, and it is common to try to escape from their contamination. Yet, if we can stay with the story, the underlying message is most likely to be rich with important guidance. Our dreams, even our bad dreams, present an opportunity to know ourselves at a deeper level. Our nightmares bring us face to face with the figurative infertility that is part of the human condition. Everyone is infertile in some way at various times.

One of the many reasons that infertility is a nightmare is because it can take a long time to come to resolution. But the time will go by no matter how it is spent. There is much to be learned by observing who we become when faced with adversity. There is opportunity in nightmarish experiences. As my colleague, Dr. Stephen Gilligan would say, "You can lose your mind, but come to your senses."

Potential actualizes when the seeds of self-awareness take root. Without self-awareness, emotional growth can be haphazard, undirected, or nonexistent. Lack of mindful awareness can result in the loss of opportunity.

## My Nightmare/My Opportunity

A nightmare that resulted in confronting adversity in my life had to do with my experiences as an ob-gyn patient. I was a young mother with a two-year-old daughter and fourteen weeks pregnant with my second child when I started to cramp and soon began to spot. I called my doctor, and he advised me to get into bed and rest. Within a few hours, however, I was bleeding. When I called my doctor again, he told me to come in to his office. When I arrived, he examined me and told me that I had just miscarried. The fetus had been expelled from my uterus but was in my vagina. He put it in a medical vial, put the vial in a brown paper lunch bag, and gave it to me! He said, "Meet me at the hospital tonight at eight o'clock for a D and C."[1]

I was stunned. The paper bag contained my baby, a baby that I had just hours before felt fluttering inside of me. I walked out of his office, dazed and speechless, not knowing how to comprehend that my child was in my coat pocket.

The whole day I was in a stupor, walking back and forth in front of the coat closet. I was inclined to open the jar and look at my baby in order to know what I was grieving for. All the while, my two-year-old was running around, reminding me of my pregnancy with her. Her sibling was in the closet in a bag.

Had this baby been another girl or a boy? Recently I saw the Bodies exhibit at the South Street Seaport Museum in New York City, which included fetuses at every stage of development. I remembered that my inclination to open the jar would at least have answered the question of gender. Part of me felt that it would have been sacrilegious to open the bag and its contents. I was not able to think clearly. In fact, I was not able to think, period. I continued to pace back and forth, robotically taking care of my daughter and arranging for someone to stay with her when I went for the D and C that night. When the time came to leave for the hospital, I had spent the day in a state of suspended animation, having never opened the jar.

When I went back for my two-week checkup the first words out of the doctor's mouth were, "How are you—fit as a fiddle?" I asked had I lost a girl or a boy. He said, "Products of conception." "Fit as a fiddle" and "products of conception" threw me back into a state of silent disbelief, similar to how I felt when he had handed me my baby in a paper bag two weeks earlier.

At that time, my level of emotional awareness and development and sense of entitlement was not where it needed to be to confront such insensitivity honestly and effectively. The day I miscarried and the day I went for my checkup, the doctor was spared from what would have flown out of my mouth if I were in my childbearing years now. How could I be fit as a fiddle when I had just lost a wanted baby? To him, had I merely lost products of conception? What the hell?

My mind froze, and the visit was merely a way of finding out that physically I was intact. Emotionally, the miscarriage felt even worse because of the doctor's complete lack of awareness. He didn't get that the pursuit of pregnancy, pregnancy itself, the childbirth experience, and for that matter, the whole issue of living in a female body at any stage of life, is much more than the functioning of a uterus, ovaries, fallopian tubes, breasts, and a pituitary gland.

The wonder and privilege, as well as the complexity, of being a woman, always part of my consciousness, was beginning to loom even larger. After my miscarriage, my emotions sent my hormones into a tailspin and vice versa. The time between my periods stretched to sixty days. The three times that I ovulated after the miscarriage before I conceived again taught me how every day can seem like a year. I was developing a keen awareness of the mind/body connection.

I went on to have two more daughters. While the profundity of this miscarriage did not render me dysfunctional, it was not something that I could easily dismiss either. It continued to bother me. It felt unfinished.

Many years went by and finally I decided that it was time to put this episode to a peaceful rest. I was not his patient anymore, but I called the doctor. I wanted the lab report. He did not have it, but if I were to write to the hospital, they would send it to me. I did. When I ripped open the envelope a few weeks later, the report revealed nothing more than—you guessed it—products of conception.

Now it was up to me to work this through. I created a ceremony in which I did a meditation that allowed me to capture what I would have wanted this child to know about how I felt about the missed opportunity to know and love him. I put my thoughts in a letter. I had intuitively felt that the baby was a boy, so I named him. I had done all of this in the comfort and solitude of my den, where a crackling, cheery fire in the fireplace was soothing me. I surrounded myself with candles, and I read the letter aloud to what I believe to be the baby's spirit. I then put the letter in the fire, giving the whole experience up to the universe. I cried and felt cleansed by my tears. It was finally over in a better way.[2]

## Another Nightmare, Another Opportunity

It took a second ob-gyn nightmare in order for me to begin to understand the huge potential to convert adversity into a new life path. The first few minutes after my last baby was born would prove to create a mystery that would take more than four agonizing years to resolve.

After Robin was born, I abruptly ceased having the level of sexual pleasure by which I knew myself. My body was not able to achieve anything more than a small ripple that was supposed to pass for an orgasm. Giving birth had been the line of demarcation. Nothing else had changed. I had always enjoyed being female and was very clear that I knew my body and its capacities very well. What was going on?

Within weeks, I went back to my doctor and asked him what could have happened. All that was different was that I had given birth to another baby. Childbirth had been a source of joy and fulfillment to me. I had my three babies naturally—no anesthesia—and I felt privileged to bring new life into the world. In fact, I loved being a woman. I wondered about the episiotomy, the surgical incision that created a larger space for the baby to come through, but the doctor denied that could be the problem. I had had episiotomies with the other two births and I was, if anything, more connected to my womanhood and more fulfilled sexually after each birth. I accepted his dismissal of the episiotomy as the culprit, but I bristled at his dismissal of me. "You are fine. It must be in your head." The room began to spin.

Over the next four years, I visited a series of other gynecologists and a few therapists. Nothing changed except my state of mind. Each new doctor who told me it was all in my head—including one female—drained a little more of the life force out of me. As time went by, I came closer and closer to thinking I was on the brink of a nervous breakdown.

Finally, a therapist told me about a segment in a book she had just read that described my complaints. The book was *Vaginal Politics* by Ellen Frankfort.[3] By this time, I was feeling quite debilitated, but the strong and healthy part of me was in a near frenzy. I ran at the speed of light and bought the book.

I opened it automatically to the exact page where, in fact, the author was describing me. In front of my eyes was the story of a woman who had ceased to have pleasure in sex after her fifth child was born. The problem was resolved when her doctor redid her episiotomy. Hallelujah! I knew it! There was hope!

I made an appointment with the chairman of ob-gyn at a teaching hospital and arrived with the book under my arm. By now, I was on fire. I actually had the nerve to say to him, "Don't you dare touch me until you read this." He read the few pages of Ms. Frankfort's book and then became the first doctor to show me enough respect to say to me, "What makes you think this is you?" I told him. When he examined me he said, "Helen, how did they miss this? I can feel that your doctor did not reattach the vaginal muscle. We can redo the episiotomy, and from what you are telling me about your history, I bet that you will be fine."

I had the surgery, and I was absolutely fine. The vaginal muscle must be intact in order to build an orgasm. The doctor who had delivered my baby stitched the surface tissue together without restoring the vaginal muscle to its proper circularity. The repair should have taken twenty minutes. In my case, the doctor had to find the vaginal muscles, which had retracted over the four years since my last daughter was born. The surgery took ninety minutes.

Good grief! If I had had less spunk, less determination, or less original connection to my femininity to begin with, it's unlikely that this would have come to a satisfactory conclusion. Women can sometimes be passive, obedient, and quick to accept blame. I met one such woman just before they wheeled me into surgery for my repair.

The woman was the nurse who did all of my pre-op checks. Although she didn't say a word, there was something about her demeanor that alerted me to her interest in my case. I told her that I had a hunch that she was interested in my forthcoming surgery. She waffled. But I could tell that she was stuck between a need to talk to me and a need to remain professional. I didn't give up. With very little additional encouragement, she burst into tears and told me that she felt the same thing had happened to her after her last child was born. She said her husband told her she "felt like an old woman to him." I asked her to check in with me after my surgery, so I could reassure her that there was hope.

The nurse's husband could not have known much about female physiology because if he had, he would have known that with age, a woman's vagina tightens. What he felt was the exaggerated openness of a severed vaginal muscle that her doctor had not properly repaired after his wife's episiotomy. There can be only limited pleasure for a man if the fit is not snug, and the fit cannot be snug unless the vagina is intact. For the nurse, the lack of pleasure was twofold—no capacity for orgasm and the pain of her husband's insensitivity.

## The Evolution of Self-Awareness

By the time I had my repair, I was a second-year graduate student in a Master of Social Work program. During the eight years in which I had been home with my babies, I had soul-searched. What would give my life meaning? Where could I derive pleasure and satisfaction from serving humanity in some way? I had enjoyed, but did not feel a passion to return to teaching fifth grade.

I had learned from my own experience and the experiences of some of my friends that there was a great need for an interface between obstetric/gynecologic care and mental health services. This was so because any ob-gyn issue is emotionally charged, separate and apart from the possibility of insensitive treatment. For some women, a simple pelvic exam is an ordeal. For others, premenstrual syndrome or menopause are devastating. The quest for a family is certainly not always the smooth path we would like it to be. And infertility may be the worst ordeal of all.

Ultimately, my experience as an ob-gyn patient was the fertile ground that led me to a profession that has been as fulfilling as anyone could dream. Those years that had been an emotional conflagration, cracked me open as if I were a lodge-pole pine seed. My true self became clear to me, and I saw to it that it would take root. I had developed the passion and compassion to help people with their reproductive imperative.

It is clear to me that I would not have happened upon my life path randomly. Tossed about by fate, my self-awareness grew. I was able to transform my nightmarish conflagration into many blessings.

For me, nurturing self-awareness resulted in my mind and body making friends with one another after years of being out of touch. Furthermore, I knew somewhere in the back of my consciousness that I had not been on an authentic path. I followed my intuition that a mind/body orientation was central to healing, and I trained in mind/body medicine. Clinical hypnosis training provided further entry into that inner sanctum where mind and body meet. My nightmares launched a process of reclaiming a resonance with my true self. My personal and professional growth unfolded from the vicissitudes and adversities of life, especially life as a woman. My experience in this realm is what qualifies me to write a book that promises to open a path to enhanced growth and satisfaction for you in the midst of this difficult time.

# Introduction

Infertility is a minefield. You never know when the next step will feel like a grenade exploded, leaving you feeling ripped limb from limb. If you are in the midst of this challenge, you understand the validity of this statement very well, but it may sound overly dramatic if you are reading this book to comprehend what an infertile friend, loved one, or patient is experiencing. Sorry to say, it is not overly dramatic at all.

In more than three decades of working with infertility patients, it was always with pleasure that I noticed how much relief they expressed just because they felt *understood.* And beyond that, relief burgeoned even more when they came to *understand* that there were stress-reduction techniques that would put them back in control of their lives. I've worked with individuals, couples, and groups. While groups were not embraced by everyone, those who joined felt as if they had found a place where they belonged, adding the benefit of eliminating the terrible feeling of isolation that is common to infertility.

The dedication that I feel to serving the needs of the infertile population had always been there. As I've acquired mind/body skills, my satisfaction has continued to increase because I am serving people in a more effective way. This alone does not explain why I wrote *On Fertile Ground.* My personal experience with the challenge of being female, and the way that my strength grew from it, made me suspect that in the advent and in the aftermath of infertility, men, women, and couples could feel changed for the better.

I set out to test my hypothesis because if I was right, I could present good news to the infertile world, beyond what was in the literature, about the stunning and ever-evolving medical approaches to conceiving. What if you could discover that the infertility struggle may seem as hard as the shell of a coconut, but, like a coconut, has sweet and nourishing water, milk, and meat on the inside?

I designed a questionnaire and sent it, with an invitation for an interview to a select group of former patients. The two dozen women with whom I spoke, and the valuable e-mail commentary that I received from several more, together with what thousands of patients had been telling me over the years, confirmed my hunch.

I asked questions like How is your life *better* because you struggled with infertility? You would think that this would raise the hair on the back of their necks. Quite the contrary. As they thought back since the resolution of their time of crisis, my interviewees came to realize the myriad ways that, in fact, many important transformations had taken place. The benefits were vast. Maturity as a person and as couples raised levels of self-esteem. People felt stronger, clearer, more connected, more authentic, and more qualified to cope with life's twists and turns. Their responses shaped this book.

I want *you* to feel understood, and I want *you* to have a different way of understanding this adversity. *On Fertile Ground* gives you the map and shows you how to navigate the territory of infertility. At the end of each chapter, you will have an opportunity to learn coping mechanisms that may not be part of your personal resources now but that can make a big difference, leaving you feeling empowered if you invest the time to learn and practice them. You will see that you can choose to turn knee-jerk *reactions* into chosen *responses* using mind/body interventions. Your ability to respond can be your *response-ability*.

In the end, you get to have the tools that give you the stamina to deal with this crisis *and* hopefully the family you long for as well. As important, when other challenges inevitably come your way, you can take solace in knowing that the mind/body coping skills that you learn now will be available to you then.

Guiding people like you to your goal of parenthood with this mind/body program and seeing your transformation in the process has been my passion. The thrill of watching these double victories is what mandated that I write *On Fertile Ground*. This book is about growing from the adversity of infertility. When faced with adversity, the normal reflex is to want to run away. Yet when we run from adversity, we run from ourselves. *On Fertile Ground* calls attention to a counterintuitive fact that if you face adversity head-on, you will find yourself on fertile ground where seeds of opportunity can take root. Your microscopic germ cells, the sperm and the egg, may not be germinating, causing great suffering, but the rest of you can be pregnant with the possible.

What if the unthinkable happens, and you are in the minority of aspiring parents whose dream does not come true? While you need not and should not be dwelling on this now, you wouldn't be normal if the thought did not cross your mind. *On Fertile Ground* invites you to cross that unthinkable bridge only when you get to it by keeping yourself invested in the present moment. Suffice it to say, for now, that growth from the decision to live child free could be an option, too. I'm going to leave this consideration until the end of the book, when what I have to say will make more sense. The main point here is that the stories contained in this book reflect arduous journeys to parenthood *and* growth from adversity. Make no mistake—growing from adversity in and of itself is a profound victory.

Meanwhile, let's get back to the task at hand, namely your intent to invest your heart and soul, not to mention your money, in the quest to complete your family. As such, you are best off knowing that feeling out of control is virtually universal for those struggling to conceive. *On Fertile Ground* will guide you to plan ways of confronting your infertility, so that, believe it or not, you will ultimately be able to find benefit in having been infertile. You will learn how to be in the driver's seat in spite of the out-of-control feeling.

This book will encourage you to step back to get a panoramic view of what happens to you in the face of this adversity. It will also encourage you to enter into the swirl of your emotions. I call the twin vantage point from which emotional growth evolves "seeing you and being you."

The orientation of *On Fertile Ground* springs from a respect for the indisputable unity of mind and body. When you can accept the reality and grieve the loss of your dream, you can then develop or deepen mind/body awareness, and finally you can put the adaptive coping skills that I provide at the end of each chapter, in place.

Infertility is a blow to the expected order of things. If the diagnosis doesn't shoot your self-esteem full of holes, the demands of medical treatment probably will. The most stress-hardy women have told me some version of "I've always been good at coping, but this is bigger than I am." The impact of infertility on any individual or couple is gigantic.

*The battleground* is the term most commonly used to label the infertility challenge by people going through it. My experience has taught me that it is a battle that all of you can come through, if not by one method, then by another. The important thing right now is feeling supported and fortified for the battle. It's like the adage How do you eat an elephant? The answer is one bite at a time.

This book strives to keep you open to what's possible. You may feel oppressed by the darkness that encroaches many times along the way, but you never know until you're looking back, whether the dark time was just before the light of dawn. People are sometimes hesitant to think positively, but the truth is that whether your stance is optimistic or not, you cannot protect yourself from the pain of a cycle that doesn't come to fruition.

Infertility creates a frenzy that sends the mind into outer space and leaves your body to walk the planet like a zombie. This book is an invitation to reenter your body, so to speak. If you feel as if you are ready to give up, *On Fertile Ground* may help to revitalize your quest and fortify your stamina with inspiration, guidance, and mind/body interventions. It is important to distinguish between a readiness to end the quest and a regrouping and going on so that you will not have to live with regrets.

Whether you've just set forth on this arduous path or you are a "veteran" and no matter your diagnosis or choice of treatment, *On Fertile Ground* will provide relief, inspiration, and hope in many ways. I'm rooting for you!

Chapter I

# Take Heart: Let's Start with the Good News

So, let's get the good news up front: You have the power to alleviate the worry about negative expectations. Scientific studies show that mind/body interventions, that you can learn, can improve rates of fertilization.

The October 2009 issue of the highly respected journal of the American Society of Reproductive Medicine, *Fertility and Sterility*, included the article *"Letting go coping is associated with successful IVF treatment outcome,"*[4] which reported the findings of a study to determine the differences among coping techniques. The researchers categorized two different styles of coping: problem-focused coping (PFC) and emotion-focused coping (EFC). They hypothesized that PFC would be more adaptive where the situations are controllable, and EFC would have more merit in uncontrollable situations such as most stages of IVF treatment. EFC was described as "letting-go coping" or "behavioral disengagement."

What is *letting-go coping* as it pertains to infertility? It can be any activity where you distract or disengage from the frenzied feeling that goes with the infertility territory. Sometimes, simple things such as absorption in a crossword puzzle can break the spasm of mind/body stress. But beyond such random activities, it is important for you to know that a purposeful absorption in letting-go coping techniques like the relaxation response, guided imagery, meditation, hypnosis, or self-hypnosis neutralize the physiology of bodily stress and have *"a statistically significant correlation with pregnancy."*[5,6] You will have an opportunity to learn many of these techniques. And, by the way, you need not be involved in in vitro fertilization to derive the benefit of this behavioral disengagement.

There's more good news. Another study, also reported in *Fertility and Sterility* (May 2006), corroborates the notion that letting go by way of hypnosis can influence pregnancy rates.

The only difference between the experimental and control groups was that at the time of embryo transfer, the experimental group received a hypnotic intervention. This study noted that *pregnancy rates of the experimental group were double* those of the controls.[7] Hypnosis is a letting-go coping method. For those of you who may not know, hypnosis does not put you under someone's "spell." Rather, is a user-friendly, natural phenomenon that among other things, typically brings deep relaxation to the person. You will have the opportunity to learn self-hypnosis at the end of chapter 7. The long and the short of it is that hypnosis or other letting-go techniques matter! No matter how you resolve your infertility, letting-go techniques can smooth the path by neutralizing stress.

Meanwhile, having and understanding information about a situation is a well-known anxiety reducer. I have watched people in your situation gobble up information and develop a proficiency in reproductive physiology that is quite stunning. I often joke with patients that with a little more coaching, they will be ready to sit for their qualifying boards in reproductive endocrinology. There is a need to make sense out of the shock of what usually feels like an improbable and unimaginable diagnosis. You can find plenty of information in books or on the Internet that tell you what to *do*. Doing is about problem solving. While this kind of information goes a long way to providing relief, it is not enough. Letting-go coping is about *being*, and I'm pleased that you will be able to learn to disengage from the all-consuming demands of infertility and just let yourself be.

But first, let's get inside the infertility experience.

## Feeling Understood and Understanding Infertility's Impact

Procreation is an imperative in nature and one from which few human beings feel exempt. Whether you are in your twenties, thirties, or forties, you think the time for creating the next generation is now. Questionable fertility, whether it comes in the form of difficulty in conception or interruption in carrying a pregnancy to term, is a crisis of biblical proportions. No one is prepared for it. Almost everyone faced with the bad news flies off into a state of bio-psycho-socio-spiritual agitation—an inescapable agitation because at your phase of life, wherever you turn, you are reminded of your circumstances. When faced with a red light to something as fundamental to existence as reproduction, the way you react emotionally will gather momentum and understandably leave you feeling ornery. It does not take long for certain personality traits that we all have to expand like a sponge in water and become uncomfortable either to you, to others, or both. Typically, infertility patients struggle with negativity, anger, lowered self-esteem, impatience, insecurity, and fear. So by now, not only has your life become unrecognizable, but if these traits have become exaggerated, you may not recognize yourself either. Trust me—you're in good company.

As time goes by without the longed-for outcome, the psychic space that you have been holding for your prospective little one looms large and is as empty as a vacuum. And, like a vacuum, it has a force—a force that sucks the existential ennui that lies dormant in all of us from its hiding place and breaks your heart. Feelings of depression are a normal response. Predictable and unpredictable aftershocks magnify the logical disorientation and uncertainty with a vibration that can only be called anxiety. The mental and physical experience of being *you* under these circumstances can be excruciating, particularly if depression and/or anxiety were unresolved companions at any prior time in your life.

To complicate matters, it's not just you in this quagmire. This may be the first time in your history as a couple that you are faced with an inability to feel in control of your destiny. Each of you may feel a seemingly unshakeable disruption in your mood, possibly in ways that may feel incompatible. Your typical coping mechanisms, and those of your partner, may be radically different. The last thing that you need right now is to feel disconnected from each other, yet this can happen.

When it comes to infertility, the complications can have complications. You love each other and want nothing more than a baby who is a product of that love. What if you are the fertile one? Think of how understandable it would be if you were to turn up angry or disappointed at the person you love the most in the world. What do you do with these feelings?

Then there is the insidious possibility that one or both of you may be blocked from knowing what you feel. Then how do you express what's in your heart? An inability to accurately understand yourself or each other can complicate circumstances. Think of how this would hamper the teamwork that makes it much easier to get through this challenge. You need each other at the same time that you may be communicating with rawness and defensiveness. Because of this, teammates can feel like enemies.

This book is also for those striving for single parenthood, and even in that case you may have someone who is supporting you who may not be exactly what you need at all times. The need to communicate clearly holds true in all circumstances. This concept also pertains to same-sex couples.

Life challenges all of us to ripen into the next phase of life. When we meet obstacles along the way, we are best off if we can search within ourselves for our greater capacities. You can take heart that those capacities are in there. Yet, when we are overwrought or angry, we could care less about reaching for inner resources, because it takes work for which we are not in the mood. How we might benefit feels squeezed out by the demands of the adversity itself. It can feel like too much to tackle while feeling so debilitated. *On Fertile Ground* will not only tune you into capacities that you can learn but will also tune you into capacities you may not know that you already have.

You're stressed to the max. Nevertheless, the ideas and tools presented in *On Fertile Ground* have the potential to take you far in your efforts to transform adversity into accomplishment. This book recognizes the difficulty of change and offers insight and guidance into the ways that it is possible to defy the influence of our emotional history and open up a path to integrating new behaviors and new ways of thinking.

## Let's Call It the Bodymind[8]

Infertility is an ultimate mind/body experience, and *On Fertile Ground* offers mind/body approaches to facing and growing from this adversity. Tackling the challenge of infertility from the perspective of the communication between the mind and the body respects the fact that the *bodymind* is a unit. We are hard-wired, so that what is felt in the body is experienced in the mind and vice versa. Under duress, our minds are at risk for flying off into the ionosphere even as our bodies walk the earth. Mind/body interventions emphasize the integrity of the bodymind even as the frenzy of infertility threatens to disrupt it at every turn. It is the mind/body orientation of this book that fosters healing and explains the subtitle, *Healing Infertility.*

It is a gross understatement to say that brain function is inordinately complicated. Yet, in order for you to come to value mind/body interventions, I'm offering a greatly oversimplified explanation of brain functioning:

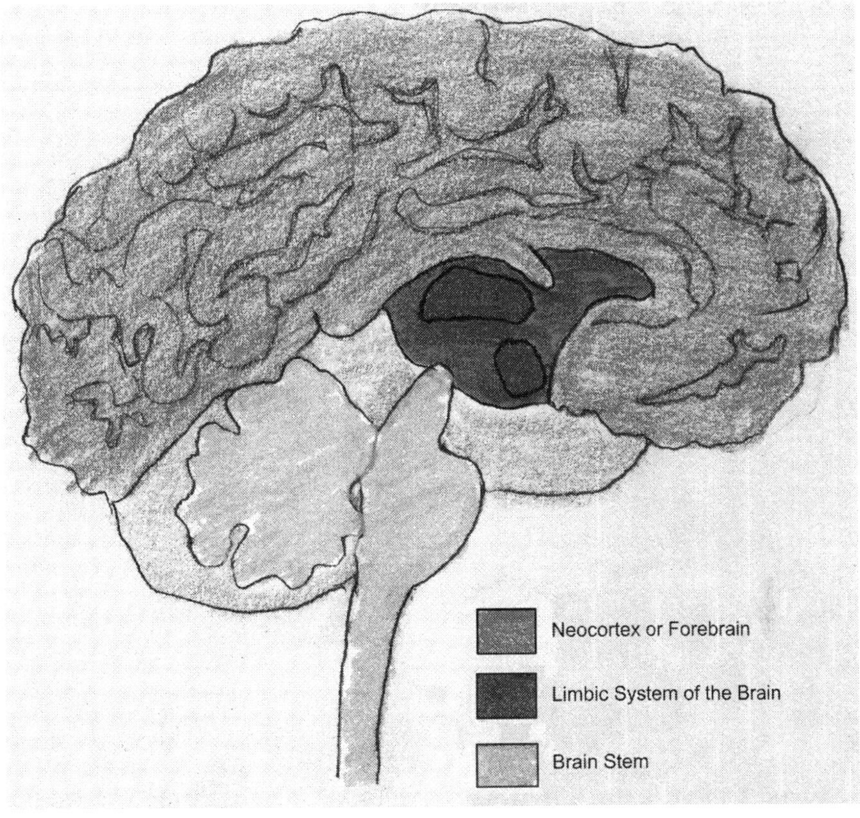


- The *brain stem* controls automatic functions such as breathing, heart rate, blood pressure, digestion, and muscle tension. These *bodily* functions elevate when we are upset (with the exception of digestion, which decreases) in preparation to deal with danger.
- The *forebrain,* includes many structures that control language, vision, hearing, planning, and much more. The neocortex, also called the forebrain, allow us to experience our world with awareness.
- The *limbic system of the brain* is part of the large forebrain. It is here that our feelings and memories are stored and here that one of its structures, the amygdala, holds large and small fears and traumas of yesterday or today even those you may not remember. Learning the "language" of the limbic system provides the best chance to find relief.

The limbic system is our inner historian of trauma and as such is a hair trigger waiting to be tripped, and the life-altering, fear-conditioning aspects of infertility are particularly good at tripping it. Limbic structures "learn" what is going on in your life from information they receive from the brain stem and the forebrain. For instance, if your eyes are closed, and you're breathing easily and regularly, as if all is right with the world, your brain stem signals the limbic system that it is safe to relax. If, simultaneously you're imagining (using your neocortex to daydream) that you're somewhere out in nature, let's say at a beach, preoccupied with imagining pleasant, sensual experiences of the sight, sound, smell, taste, and touch of the environment, *the limbic system of the brain does not know that you are not where you imagine yourself to be.* It responds as if you are there, and the bodymind can let go of the physiology of stress. You will learn several letting-go coping strategies that will enable you to "communicate" with the limbic system.

When it comes to using mind/body techniques, there is the linear left-brain functioning and the creative right-brain functioning to take into consideration. The limbic brain understands creative communication, therefore it prefers stories about change rather than lectures about how or why to change.

So much for the brain. Then there's the mind. Let me simplify another complicated concept. We have a conscious mind that is aware of what it's aware of. But we also experience and store information of which we have no conscious awareness. Of greater magnitude, the unconscious mind is always conscious even when the rest of us is in la-la land. Letting-go skills involve an invitation for the conscious mind to step back as observer giving the unconscious leeway to participate. While we can tap into unconscious wisdom with letting-go coping, part of unconscious functioning can also put obstacles in our path. You will learn more about how to circumvent this pitfall in chapters 6 and 9.

By participating in your own care with the mind/body techniques presented here, you can feel as if you are a partner to yourself, a partner to your partner, and a partner to your doctor. In essence, you can ease the pain of the circumstances by regaining the feeling of being in the driver's seat of your life, even as infertility seems to toss you into the bushes.

## The Heart of the Matter

The diagnosis and treatment of infertility are science and technology for the body. *On Fertile Ground* addresses the heart. Longing for a child is the heart of the matter, yet western medicine's orientation isolates the body parts that are functioning inefficiently. It is a disease model (which means that the focus is on what's wrong in hopes of fixing it). My orientation is toward the totality of your mind/body/spirit, and the reality of the social environment in which your infertility struggle takes place. This orientation goes beyond holistic in the sense that mind/body integrity is more than the sum of its parts. This book invites respect for this totality while at the same time acknowledges respect for the miracles that science and technology can create. This book is about balancing the science and technology by bringing all of you, especially your heart, into the story.

Time will pass no matter how you are spending it. This book makes a case for using this time to study yourself in these unwanted circumstances. To simultaneously *see* you and *be* you, even as the various assaults of the fertility struggle descend on you, gives you your best shot at engaging in the process, resulting in a sense of satisfaction, accomplishment, and control. This book can put you in control of how you navigate your journey, even if you cannot control the fact that you are on it. This may seem like a small difference, but it is the difference that can make all the difference.

## Triumph

Learning the mind/body program represents a conception of a different kind. Awareness is integral to healing. You have an opportunity to "conceive" of yourself in new ways. The sudden realization that your goal to enhance your family has been thwarted ambushed you. The last thing you would think is, "Oh, goodie, I'll be a better person when this is over." If you're normal, what you really want is magic. Yet I can say unequivocally that the byproducts of a shift in outlook can result in a great triumph while you wait for your family to include little ones. It is possible to go beyond surviving the challenge by learning how to thrive in spite of it.

Andrea was an infertility patient of a colleague of mine. She was kind enough to read an early draft of this book and she captured my central message:

> I was struck by the thought that turning inward (learning to *be*) not only helps to provide a diversion for the infertility issue, but it also gets deeper into long-term healing. I was thinking that at such a turbulent time, it is almost overwhelming to think of examining who you are; yet if you can push past it, the rewards are plenty. After reading this book, I felt very hopeful. Learning how the infertility

process became a learning experience that had lasting effects made it seem almost worthwhile. It is your choice to make it what you wish.

Soon after reading this book, Andrea conceived. You can take heart if you can anticipate feeling renewed hope the way Andrea did.

I formally interviewed more than two dozen former patients in preparation for writing this book. They chose to share the upside results of their downside experience because they wanted you to be inspired to keep the faith. These people rode the not-so-merry-go-round of what to *do* about their longing for a child, and in the process, discovered a more profound and satisfying way to *be*. It was unanimous among my interviewees that the benefits of going within themselves and rising above themselves during the infertility challenge gave them the confidence to feel that there is nothing that they cannot handle. I heard this comment so often that it was another factor that prompted me to write this book.

I encourage you to start with the following attitude rather than energize the normal worry that nothing will work: Let's presume that your child will be conceived, even if, in the final analysis, that child grows in a surrogate's belly, is the product of someone else's genetic material, or is born *for* you. The many options for greater self-awareness along the way can be gifts in disguise for their capacity to transform the infertility desert into fertile ground, full of possibilities. From this vantage point, the fertility quest, working through an unwanted adversity, is about more than getting a baby—a lot more.

# EXERCISES FOR STARTERS

**"For fast acting relief, try slowing down." —Lily Tomlin**

Each chapter contains options for coping with the unwanted change in your expectations. The exercises, techniques, or questions to ponder are designed to move you through the infertility quagmire. The main goal is to *see* you and *be* you, so that you can identify what you need to change. This book will help to free you to get obstacles to change out of the way.

We learn best when we are relaxed. An article in the science section of the *New York Times* reported that we learn more efficiently when we are asleep, and we dream about what needs to be learned.[9] Yet, when awake, stress notoriously creates what is called a fight-flight-freeze reaction. We either mobilize to deal with the stress, run from it, or freeze, feeling helpless. Our physiology follows suit, and before you know it, the autonomic nervous system takes over and heart rate, breathing rate, blood pressure, and muscle tension elevate.

### Exercise 1

While breathing is automatic, it is common for anxieties to result in shallow breathing. Since the breath is a built-in tranquilizer, you have the power to reverse the physiology of stress by consciously breathing deeply. One luxurious, slow breath can provide a *sigh of relief* and shift your physiology back toward neutral. To focus on an exaggerated but not a forced breath is a perfect introduction to the best way to *be you*. Start now by *taking a breather*, a generous inhalation and release. You should feel a shift, no matter how small. If you don't, try again.

### Exercise 2

A more elaborate exercise involving using the power of the breath to tame frenzy is called *the relaxation response*. It is an exercise that has been proven effective in many studies conducted by its pioneer, Dr. Herbert Benson of the Harvard Mind/Body Medical Institute (now called the Benson-Henry Institute for Mind/Body Medicine). It is a way to distract the conscious mind from agitation while using the breath to sooth the nervous system. This technique will keep you focused in the present moment. This is important because levels of stress are bound to increase when we ruminate over yesterday or tomorrow.

Dr. Benson's studies have been replicated and expanded for use in many areas where the physiology of stress needs to be reversed such as lowering blood pressure; there is no question that all you need for the relaxation response to be effective is to do it. It is easy to learn. While accessing the power of conscious breathing for its natural tranquilizing effects, repeat in your mind (not aloud) a word, phrase, or prayer to which you have a positive association. For instance, you can mentally repeat words from nature such as *rainbow* or *ocean*. You can say action words such as *love* or *relax*. You can use adjectives such as *peaceful*, or commands such as *be serene*. You merely need to say your choice of word to yourself every time you exhale. You're likely to lose your place. Without judgment, let it go and start over.

As an alternative, using a phrase such as *I am…at peace* or *I will…conceive* can be powerful. Here you would mentally repeat the first half of your phrase on the inhalation and the other half on the exhalation. If you feel you'd be best using a phrase from favorite prayer, such as *The Lord…is my shepherd,* say half of that on the inhalation and the other half on the exhalation. It doesn't matter what you choose. You could use your dog's name if that gives you solace.

That's it! Let's try it.

1. Make yourself comfortable.
2. Take an exaggerated breath, enjoying your capacity to bring in the life force with each breath. *Keep in mind that the **life** force is what this is all about.* Enjoy this inhalation and exhalation as a way of separating *before* from *now*.
3. Scan your body, noticing places in your body that are at ease and places where you might be holding tension. As you are resting in the rhythm of your breathing, trust that you can expand the places where you feel comfortable, superimposing the calm feeling on the places where tension has taken up residence. Let your body soften to its potential to let go and let down.
4. Select a word, phrase, or prayer. Gently begin to repeat your choice of your silent chant, coordinating it with your breath as directed.

5. Stay in this mode for ten to twenty minutes, keeping in mind that *when* you lose your place (everybody does), just let it go and start over.
6. Bring a nonjudgmental attitude to this exercise. It may be difficult for you to stay focused for twenty seconds, never mind twenty minutes. This is normal. With practice, your body will acclimate to a new set of rules. If you tend to be the type of person who rarely sits still, it may take longer for your body to realize the value of this gentle way of taking care of yourself.
7. At the end of twenty minutes (you might want to set a timer), you may be amazed that a dramatic shift has taken place. Enjoy the amazement. The reason for this will become clear later in this book, but for now, acclimate yourself to the process.
8. Many people have trouble maintaining focus at first. This is a skill, and like any skill, takes practice. Hang in there. I'm not selling snake oil.

For additional help with this technique, click on the "Purchase" tab on my Web site, www.mind-body-unity.com, to purchase a CD that supports the practice of the relaxation response.

Chapter 2

# Infertility's Impact on You: Turn the Tables on Change

Life has changed. Supply your own expletive here (______________!#%&!).

The change might be everything that goes along with the shock of the diagnosis. But if you are in the roughly 18 percent of couples whose doctors are unable to provide them with a diagnosis, it may take longer to realize that you've gone from being a "normal" couple to one who is in the minority. No matter what, life has changed.

This chapter talks about change in general, changes particular to the fertility struggle, the emotional impact of these changes on *you*, and how to begin to achieve a mind/body perspective that will launch you on the path to feeling better.

Any change is uncomfortable because you're in unfamiliar territory. Yet, change is a fact of life; it is inevitable and nonnegotiable. What makes infertility so hard is that it's not a natural change like graduating from high school or getting married. Infertility is a change that has a psychic violence to it. It is an abrupt blast of unwanted reality. When something ends without a warning and is replaced by something unwanted, the discomfort can be agonizing. No wonder the sense of being in the driver's seat of your life evaporates.

When Donna, an editor, learned about her infertility, she remarked that it was as if someone pulled the rug out from under her. "Once you get slapped in the face with the diagnosis of infertile, no matter how smart you are, no matter how pretty you are, no matter what you have, it just doesn't matter...It's completely out of your control," Donna said. Once you understand that the feeling of losing control is part of the infertility picture, you can understand that among the infertile population, you are normal.

Then there's the stress engendered in this change. Stress and infertility are travelling companions. They are thick as thieves. They move into the center of your soul together, and they are like unwanted, annoying guests who act like they own the place. The diagnosis is stressful, the treatments are stressful, and let's not forget that you probably have a job and a life, which could be fraught with stressors of their own.

These gut-wrenching intruders, stress and infertility, can sap your energy physically, emotionally, socially, and spiritually, and unless you get lucky and conceive quickly, you're bound to get smacked in the head by this devilish duo.

You have every right to feel devastated. Studies have shown that levels of stress that come with the diagnosis of infertility are on a par with the diagnosis of cancer, heart disease, or HIV/AIDS.[10] Hear this, because it is important: *No one should consider him or herself too upset.*

It's normal to feel out of control, normal to feel inordinately stressed, and it is also normal to resist facing the truth. It is common to protect yourself by denying that this profound change has really happened. While this can be true of all kinds of change, resistance and denial are even more logical when the change blindsides us. And why wouldn't there be resistance and denial? You need time to wrap your brain around what's happening. The first inclination is to hope this is all a big mistake. When it becomes evident that you're dealing with questionable fertility, where your dreams are shattered, it's normal to slog through a period of time before you can find a new center of gravity.

By now, you may be thinking, "Great…so I'm normal…big whoop." Even though you have a right to feel like crap, *On Fertile Ground* shows you that you have a choice not to.

> After trying to get pregnant naturally, we realized that things were not happening. I realized I was in overdrive, and *I needed to change.* I learned that I could *choose* ways to deal with infertility, so I could have hope *and not let my mind tell me otherwise.* I was working way too hard. Eventually I quit my job and took some part-time work. I focused on bringing my body into a more grounded, healthier state. The anorexia in my past made it necessary for me to reconnect with what my body was supposed to do, and I learned how to be more at peace with myself (*Italics are author's*).
> — *Julie, stay-at-home mom, Fairfield, Connecticut.*

## So, Now What?

Let's start with this: The great religions of the world teach us a key concept about devastating change. The many-armed Hindu god Shiva, for instance, is poised with one foot about to stomp on the earth. This symbol communicates the belief that it is only when something is destroyed that something better can emerge. Christian ideology tells us how Christ needed to die so that He and His followers could be resurrected. Judaism teaches that immediately before the advent of the Mashiache (the Messiah), there will be a time of great travail and turmoil. It is only into this troubled setting that the Messiah will come.

Despite the fact that the need for change has come in an abrupt, violent way, in the giant scheme of things, even if you think it sucks, it is the way of the world. You may not give two hoots about this right now, and who would blame you. However, you may find a bit of comfort in the fact that we are hard-wired to snuggle into sameness and predictability, *and* we are hard- wired to adapt to and even seek change. We are biologically qualified to meet difficult challenges, even those that provoke disturbing internal conflict and confusion.

In fact, what I have most commonly seen in my decades of work with patients is that the strength and resources that you didn't even know you had can be called forth. Sure, you may collapse from time to time. That just has to do with the need to catch your breath. Keep in mind the magnitude of the challenge you're dealing with. As challenges go, this is Mt. Everest, and who wouldn't need a way station where you could plop down and rest up for the next leg of the climb?

## The Emotional Impact on You

In the beginning of the journey, you might find yourself in shock saying, "We're okay—we'll get through this." Eventually, it is most usual to arrive at a place where the predominant feeling becomes is this ever going to end? The challenge is like a twenty-eight-day roller-coaster ride—climbing to heights, feeling cautiously optimistic, and then feeling the downward rush of disappointment when you get your period. How many months in a row can you expect yourself to maintain stamina under these circumstances when the descent saps the energy that you need for the journey? No wonder, despite the strongest resolve, experiencing a monthly "death" puts you face-to-face with the real issue: loss and grief. The emotions that go along with the loss of your dream are the antithesis of the excitement of starting a family. You may be full of longing, but at the same time, you may have become devoid of joyous anticipation. It is natural to want to avoid the painful feelings of grief and loss. But here's the problem: while your mind may be scrambling to be anywhere but in the pain, your body is always present to it. One of my patients said it this way, "I feel as if my mind and body are going in opposite directions." This is why *On Fertile Ground* brings a mind/body orientation to the solutions offered throughout this book and to the exercises at the end of each chapter, so that you can meet these mind/body challenges with mind/body integrity.

As the dust begins to settle, you can stand to regain a sense of control by realizing the importance in both seeing you and being you. Seeing you means stepping back so you can gain a perspective on how you *react* to the travails of what most people call the battleground. Being you means experiencing the way the mind and the body are screaming at each other like a seriously dysfunctional couple. You can assume response-ability by learning how to convert automatic reactions of frustration into thought-out responses.

As chaotic or even impossible as this whole thing might seem to you at this point, you will see that you can steadily learn to take responsibility for transforming grief and loss into empowerment, one little victory at a time. Empowerment is the fuel that keeps hope alive for the journey.

## Change Your Perspective: The Three As

Reactions can become responses as you become aware of your options. Options begin to flow from what I call *the three As*: *Accept* the reality, develop self-*Awareness*, and learn to *Adapt*. The three As are the Siamese triplets of healing. You'll be hard pressed to change without acceptance. From there, expanding awareness of yourself (and those around you) is central to developing the flexible thinking that you need. This will provide the motivation to learn the tools to adapt, so you can change what needs to be changed. Start now by taking a breather.

You can plant yourself in the fertile soil of acceptance where regaining control of your path toward parenthood can germinate. Awareness and adaptation can soothe what's tearing your heart out and ease the intensity of the upset by opening up new possibilities, particularly with letting-go coping techniques. This is probably hard to imagine right now, but unless you can somehow shift into acceptance and start working with awareness on adapting, you'll end up feeling as if you have one foot stuck on flypaper. Without acceptance, awareness, and adaptation, the feeling of being out of control can morph into something even worse—the feeling of helplessness.

I never pursued anything like I pursued this; I knew I was strong in a way that I knew I could handle shit. I felt that shit was all I ever had, and I learned to deal with it. But what I really learned was that I could actually deal with it in a way that was not just resigning to it, but *changing* my attitude and really not quitting.
— *Robin, antiques dealer, New York, NY*

**"To do is to be."** — *Socrates*
**"To be is to do."** —*Plato*
**"Do, be, do, be, do."** — *Sinatra, "Strangers in the Night"*

Most people find it a lot easier *to do* something when troubled by grief and loss than *to be* with these feelings. Yet the opportunity is right in front of you to delve into being aware of what your body and mind are communicating to you, thereby honoring the bodymind. *Doing*, based on action, is more concrete and easier for most people than *being*, which necessitates having an ability to tolerate what is. For example, everybody knows somebody who, when faced with a difficult emotional situation, goes into a frenzied cleaning mode to avoid dealing with his or her emotions. This person vacuums everything that can't walk away, or rips a closet apart whether it needs to be reorganized or not. Doing is what people do when they don't know how just to be with their emotional selves.

When it comes to infertility treatment, your medical team will be telling you plenty about what to do. Yes, doing what is called for can be palliative, not to mention necessary. But dealing with the emotions, the being aspect of coping, is what keeps the bodymind intact. Healing infertility has much to do with awareness of mind/body unity when it comes to dealing with the core of the infertility problem—grief and loss.

## Grief and Loss

For sure you feel a loss, but unlike the death of someone who has been alive, it may not occur to you that you are grieving. We tend to want to avoid grief, yet the way out of any emotional experience is through, not around.

In 1969, Dr. Elisabeth Kubler-Ross wrote her landmark book, *On Death and Dying*.[11] In it, she delineated five stages of grief. You are likely to relate to the list:

- *Shock and denial* (including avoidance, confusion, fear, numbness, and blame)
- *Anger* (including frustration, irritation, embarrassment, and shame)

- *Depression and detachment* (including feeling overwhelmed, the blahs, low energy, and helplessness)
- *Dialogue and bargaining* (including reaching out to others, desire to tell one's story, and struggling to find meaning in the loss)
- And finally *Acceptance* (including exploring options, making a new plan, feeling secure, feeling empowered, elevating self-esteem, and finding meaning in the loss)

Dr. Kubler-Ross laid these out as having a flow from beginning to end, but I find that there is not necessarily an order or a timetable as to how these aspects of grief and loss are experienced.

*On Fertile Ground* is designed to highlight that grieving losses is important in arriving on the other side of upsetting circumstances but is also important in gaining the ability to open up to new circumstances.

## Doing Versus Being (or Problem Solving Versus Letting Go)

The bottom line is that problem-solving coping has a place when there is a solution that matches the problem. Since the only solution that matches the problem is bringing a baby home, and since infertility is so gigantic and complicated, letting-go coping is more effective.

Medical treatment falls in the category of problem-solving coping. So do the cognitive/behavioral mind/body techniques, such as the ones at the end of this and other chapters.

Letting-go coping techniques include meditative approaches in which your focus is narrowed, allowing you to experience and ease the impact of mind and body on each other. Besides meditation per se, other approaches include guided imagery, mindfulness, breathing techniques, hypnosis, or self-hypnosis (which I will teach you at the end of chapter 7).

> **"There needs to be an infertility-free zone"** —*Amelia, drama coach, New York, NY*

Ask an infertility patient if there's an *infertility-free zone*, and most will think you've lost your noodles for even asking. Given the feelings of frenzy, no one can image that such a place exists. But it does, and it has to do with mastering letting-go coping techniques. The infertility-free zone is not a place "out there" to which we can run. It resides within each of us. *Within us,* however, is not a zone that most of us are accustomed to visiting.

Here's how letting-go coping works. It's worth repeating the workings of the brain. Remember, the brain stem controls our automatic bodily functions like heart rate, respiration rate, blood pressure, digestion, and muscle tension. The neocortex controls cognitive functioning. In between the two, the limbic system of the brain connects the mind aspect of brain (cortex) and the body aspect of brain (brain stem). The limbic structures are wired in such a way that they know what the person is experiencing based on what signals come to it from the brain stem or the neocortex. So, if you are anxious, but you breathe slowly and deeply, as far as the limbic brain is concerned, you are relaxed. Then, if you can divert worrisome thoughts by imagining yourself at a place of solace like a lake or meadow, the limbic brain enjoys the sensual pleasures of that fantasy as if you were really there. The combination of mindful breathing with pleasant visualizations can drain stress out of your bodymind.

Another important factor in developing letting-go skills is the fact that the limbic system has no concept of time. Therefore, if you spend only a few minutes in a pleasant reverie, as far as the limbic system is concerned, those few minutes can feel long enough to break the spasm of stress, leaving you feeling relieved. The limbic brain's capacity for time distortion is why doing the relaxation response can be so palliative.

Even the cognitive problem-solving techniques such as the ones at the end of this chapter, designed to build awareness and promote adaptation, can have a letting-go effect if you take solace in doing these self-rating scales and then following through on making changes. By changing something that isn't serving you, it becomes easier for you just to be. *On Fertile Ground* will open a vista where changing the internal landscape of your bodymind becomes comprehensible, desirable, feasible, and, compared to the expense of medical treatment, affordable.

## You're on Fertile Ground

The more that you embrace what infertility has the power to teach, the more you can realize that being called up short in the face of life's nastier circumstances gives you an opportunity to grow tall. The more self-aware you become, the more likely you are to direct your life toward its best unfolding. As self-awareness increases, and armed with mind/body approaches as part of your repertoire, you are set to have the tools to discover and deal with what needs to change. This is what you *can* be in control of. Regaining control is part of healing infertility.

The road to filling your crib exposes many common and predictable challenges. Embrace them, because these challenges offer you a chance to give birth to a new, improved version of yourself. The fact that these are common challenges is why fellow comrades, who have battled this same war, are sharing their stories with you in this book. You will see how, with awareness and mind/body tools, you can sculpt changes rather than be chiseled away by them. What a relief when you can feel empowered and inspired as *you* mastermind changes that ease the stress.

## Exercises

## So Let's Get Practical...

This chapter recommends that you review the four mind/body self-assessment scales that follow. They will provide an opportunity to examine your lifestyle, notice your stress warning signals, evaluate your level of self-esteem, and understand your underlying beliefs. This strategic approach gives you a chance to see you. Then you're in position to decide what has to change in order to minimize stress. Get to know yourself. Knowledge is power.

# The Lifestyle Profile

The *Lifestyle Profile* is designed to reveal where you stand in terms of social isolation, nutrition, assertiveness, exercise and recreation, sleep patterns, psychological health, and spirituality. These are important aspects of self-care. What you learn can help you to become aware of how you may inadvertently be contributing to your stress. Then it's up to you to implement changes that would relieve the stress associated with the issue at hand.

## *Lifestyle Profile*

Infertility is a bio-psycho-socio-spiritual crisis. It demands good self-care. Here is your chance to *become aware of the power that you have to make changes* in these areas that would translate to taking good care of yourself at this critical time.

Biological Considerations:

Do I work too hard?

Am I nourishing myself in a healthy way in general?

Have I researched nutritional advice that enhances fertility?

Do I know what foods/drinks to avoid?

Do I exercise?

Do I know what exercise is optimum for me under these circumstances?

Do I exercise too much or too little?

Do I balance work and play?

Do I create time to relax, optimally every day?

Do I know how to manage stress?

Do I get enough sleep?

Psychological Considerations:

Do I feel loved?

Do I feel empowered or entitled to change things?

Am I aware of my feelings?

Can I assert my needs with family, friends, and coworkers?

Do I know how to get information that will relieve my upset?

Am I interested in emotional growth and development?

Can I see beyond this crisis to an exciting future?

Have I accepted reality so that I can strive to feel empowered to regain control of my life?

Do I like who I am?

Do I react to stress or respond to it?

Can I access my creativity?

Social Considerations:

Do I feel isolated?

Have I always felt isolated?

Do I know how to get social support?

Do I want social support?

Do I know how to find social support under these circumstances?

Do I know how to care about people?

Do I let them care about me?

Do I like how I behave with others?

Spiritual Considerations:

Do I feel connected to a higher power?

Is it okay with me if I don't feel connected to a higher power?

If I've lost a spiritual connection, how can I get it back if I want to?

Do I feel punished?

Am I angry with God?

Does my faith provide solace?

Do I believe that my life has meaning?

## The Stress Warning Signals

The ***Stress Warning Signals*** scale gives you an opportunity to realize where stress has "landed" in your physical body or mind. Symptoms are the body's wisdom. Identify the symptoms as gifts from your bodymind. Knowing how the mind manifests stress in your body and how your body manifests stress in your mind gives you the information you need to target solutions.

### *Stress Warning Signals*

Physical Symptoms

___Headaches
___Back pain
___Muscle tension
___Digestive problems
___Lowered immune response
___Exhaustion

Behavioral Symptoms

___Too much or too little food
___Too much or too little sleep
___Overuse of alcohol or cigarettes
___Inability to get things done
___Excessive shopping or exercise
___Teeth grinding
___Isolation from others
___Nervous habits (nail biting, for example)

Psychological Symptoms

___Crying
___Edginess, moodiness
___Racing thoughts
___Depression
___Feeling overwhelmed
___Helplessness
___Hopelessness
___Guilt
___Anger
___Impatience
___Insecurity
___Lowered self-esteem
___Sense of fear or dread
___Ashamed
___Victimized

Cognitive Symptoms

___Difficulty concentrating
___Forgetfulness
___Memory loss
___Indecisiveness
___Loss of sense of humor
___Loss of objectivity

Relational Symptoms

___Intolerance

___Resentment

___Loneliness

___Lashing out

___Clamming up

___Isolation from friends/family

___Nagging

___Distrust

___Lowered sex drive

___Lack of intimacy

Spiritual Symptoms

___Emptiness

___Cynicism

___Unforgiving

___Loss of direction

___Apathy

___Indecisiveness

___Loss of meaning

___Looking for meaning

# The Self-Esteem Checklist

The ***Self-Esteem Checklist*** will give you an opportunity to become aware of if and how your self-esteem has faltered. Then you will be able to reclaim and fortify your sense of yourself. It is common even among those who consider themselves to have excellent and reliable coping skills, and who have coped well throughout their lives, to feel leveled by the infertility experience.

## *Self-Esteem Checklist*

1. Do you compare your inside to everyone else's outside and come up short?
2. Do you hold back on expressing your feelings?
3. Do you compete to win and feel lousy if you lose?
4. Do you feel hesitant to try new things especially if others are watching?
5. Do you tend to put others before yourself?
6. Is it hard for you to accept compliments?
7. Do you feel as if you are a victim?
8. Do you compensate for feeling *less than* by bullying others?
9. Do you hold yourself to ridiculously high standards and then judge yourself harshly if you do not meet them?
10. Do you lack confidence in your coping skills?

How many of these questions have you answered *yes* to? Can you determine whether the infertility experience has shot holes in your self-esteem, or whether your self-esteem needed a shot of B12 to begin with? Keep in mind that this book and the exercises in it are meant to help you build a stronger sense of yourself. Hang in there!

## The Beliefs Inventory

The ***Beliefs Inventory*** will unearth underlying beliefs that undermine hope. If beliefs are underlying, you may not be fully aware of the attitudes that can flow from them. However, you can learn to spare yourself from a sneak attack. Find out what these beliefs may be. Clear the path to hope.

Answer yes, no, or sometimes to the groups of questions below. **Then match your answers in section A with the explanation below in section B.**

## A – The Beliefs Inventory

I

a. Do you need family, friends, and coworkers to show approval of whatever you do?
b. Do the opinions of others influence your behavior?
c. Do you need everyone to like you?
d. Are you very sensitive to criticism?

II

a. Do you strive to be perfect?
b. Does it bother you if you feel that others are better than you?
c. Do you become annoyed if you do not succeed at something?
d. Can you compete for the fun of it?

III

a. Do you judge others harshly because you were judged harshly?
b. Do you believe that punishment keeps people in line?
c. Do you believe in getting even?
d. Do you expect a person whose response to you is upsetting to behave differently the next time you're in the same situation?

IV

a. Are you easily frustrated?
b. Do you get riled up if a situation is not what you think it should be?
c. Do you recoil at challenges?
d. Do you have trouble accepting what is?

V

a. Do you feel justified in reacting with vehemence when things don't go your way?
b. Do you believe that life should be easier?
c. Do you find it hard to be happy?
d. Do you believe that you cannot help whatever mood you're in?

VI

a. Do you believe that the world is an unsafe place?
b. Do you hate uncertainty?
c. Do you worry incessantly?
d. Do you try to control people and events to feel more comfortable?

VII

a. Do you second-guess yourself when it comes to managing your life?
b. Do you rely on others for decision making?
c. Do you resist doing unpleasant chores?
d. Do you avoid responsibility wherever you can?

VIII

a. Do you look to those in positions of authority to depend on?
b. Do you believe that others are stronger than you are?
c. Do you need to lean on others when you have problems?
d. Do you need others to be concerned about your welfare?

IX

a. Do you believe that the past predicts the future?
b. Do you believe that your history will limit your life path?
c. Do you believe that people can never change?
d. Do you fear the future?

X

a. Do you believe that happiness should just come your way?
b. Do you prefer to spend as much time as possible in leisure activities?
c. Would you resent having to work hard?
d. Would you want to organize your life, so you can work as little as possible?

## B – Understanding the *Beliefs Inventory*

Each group of four questions in section A addresses a belief that, if held, is likely to lead to dissatisfaction. Answering *yes* to more than one question in each group signals you that you have an investment in a belief that if examined, can be reworked, providing relief. You will learn at the end of chapter 5 how to reorder your unrealistic expectations, so that you are not trapped in assumptions that can be hurting your ability to cope with adversity. Meanwhile, here are the ten categories of underlying beliefs that might have taken up residence in your psyche. Match these explanations with your answers above.

I. If you were not given adequate love and approval as an infant and young child, you might be at the mercy of the opinion of others as an adult. Love and approval from one's self is the best and only rational guide.

II. The belief that it is necessary to be perfect and competent in all you do could stem from a failure to have been accepted for your best effort as a child. Your ability to accept yourself with flexibility would be compromised. It's great to have high standards and crippling to expect perfection all of the time.

III. Sad to say, some people are unworkable because they are rigid and self-involved. If you were raised by someone like this, you may feel helpless when you encounter this personality type later in life. You need to recognize people who are cruel and heartless, and set limits with them or eliminate them from your world, or you are doomed to feeling enraged and frustrated.

IV. It is unreasonable to expect things to turn out the way you want them to. If when you were growing up you didn't learn to discriminate what you can control from what you can't, then as an adult, you are primed to be disappointed.

V. It is a mistake to believe that human misery always comes from the outside environment. You can make yourself miserable from the inside. Growing up healthy depends on developing conscious awareness and learning to take responsibility for your actions.

VI. Life on earth is about uncertainty. Children need to be encouraged to face the inevitability of the new/unknown. Without encouragement, you can feel paralyzed by fear and completely miss the potential for excitement and growth.

VII. If you were not encouraged and supported in facing life's difficulties as a child, you might be inclined to avoid life's difficulties as an adult. You might think that a solution is to rely on others, but this is dependency or codependency. And, if your "solution" is avoidance, you may have noticed that things may actually get worse, since problems and responsibilities do not go away.

VIII. Perhaps you were led to believe that you always need someone stronger or smarter than you upon whom to rely. Healthy self-esteem does not preclude asking for help, but to believe that you can never rely on *your* strength is an insult and compromises your ability to esteem yourself.

IX. If you "marinated" in your parents' learned limitations and took them on as your own, you doom yourself to repeating the past. The past is a strong force, but not a life "sentence." Change is difficult but possible.

X. If you learned that happiness can be achieved by inaction, passivity, and endless leisure, then you learned to rely on magical thinking. It makes no sense to believe that happiness comes serendipitously; this can only lead to anger at the universe for failing you.

Stress is something that can come from within us as well as be external to us. These self-rating scales will reveal information to you and about you, and reveal where you can make lifestyle modifications. This will enhance your feeling of empowerment—all the better to make it through this endurance test to parenthood.

Chapter 3

# Infertility's Impact on Your Relationship: Maintaining Intimacy

Under the best of circumstances, the context in which this crisis hits is unfortunate. Chances are you are in a young marriage/relationship. Your capacity to navigate rough times together may be unpracticed. If your relationship has longevity, you might have been lucky to have had a relatively smooth sail until now. Or perhaps you have forged through other crises of lesser magnitude together, where the strength of your respective mental muscle enabled you to muddle through while remaining balanced enough to cope as teammates, even if clumsily.

Infertility may be a bane to your existence, but in important ways, in the final analysis, it can be a boon. If it doesn't drive a wedge between you and your partner or even if it does, the challenges can ultimately solidify your love for one another, not unlike soldiers who bond for a lifetime as a result of sharing the dangers of battle. This chapter will give you ways to enhance your teamwork and keep your heart in the story.

At the outset, couples usually vow to use the challenge to develop a closer relationship. But, if the months have already dragged into years, it may seem daunting to sustain momentum despite your best intentions. Do you *each* have the skills, the shock absorbers, and the stamina to endure this potentially long journey? What if your coping styles seem incompatible? Do you have experience dealing with the tough stuff as a team? Have you been able to work things through in the past even if you haven't agreed on important issues?

If these questions leave you feeling a bit queasy, all is not lost. The stress that comes with infertility packs a predictable wallop from many fronts, and it's hard enough to keep the lines of communication open when things are going great. If things have spiraled out of control, you're in good company. Unless your elders were great role models, you may not know what you need to know to ride this challenge to a satisfactory conclusion. If after reading this chapter you still feel unsure that you can learn to tackle your infertility crisis as a united front, get help—fast.

Many couples report that after they get through the infertility nightmare together, they feel closer and may already have proven that despite being tested, they're solid. Adversity has the potential to pry open a panorama of infinite possibilities in a *good* way. Empathy and understanding can deepen and flourish. For example, in the case of Brian, now a stay-at-home dad, when he and his wife were battling infertility, he came to realize that "What happens to Stacie happens to me." Smart guy.

Respect can grow. During their infertility crisis, both Julie's and Lauren's husbands, Dan and Kevin, told their wives how much they admired the way they were enduring the seemingly relentless assaults of treatments and disappointments. Potentially, loving feelings of all kinds can come into full bloom when you bond as comrades in what Monique, an MBA holder and lawyer by training, called "the trenches." It is heart warming when partners do not take each other for granted.

*The primary ingredient to getting through this heart-wrenching time is awareness.* It is optimum if self-awareness is balanced with understanding how your partner is feeling. I call this *other-awareness.* Remember, awareness is about entering the place where mind and body meet. Sufficient awareness, self-care, and caring about each other can result in mental and physical well-being.

## Resist Playing the Blame Game

If there is not a history of self-reflection and empathy in your relationship, then the stress from the infertility battle could lead to finger pointing and accusations. Blaming each other can cause negative feelings to set like cement. That would ultimately prevent the teamwork you need.

There is a common trend to see our partner's liabilities clearly, while remaining sketchy about, if not oblivious to, our own. I call this the If-it-weren't-for-him/if-it-weren't-for-her-we'd-have-the-best-relationship syndrome.

Under these circumstances, We'd-have-the-best-relationship-if-only… is a wish based on a storybook longing. If earlier circumstances in your life prevented you or your partner from learning how to understand and tend to each other, then expecting an ideal relationship would be like expecting that you could fly to the moon in a rowboat.

**CRISIS AS OPPORTUNITY**

The Chinese word for crisis is comprised of two parts: *danger* and *opportunity*. Danger is "pictured" as a man on the edge of a precipice; the symbol for opportunity is meant to be a reminder that something positive can come out of danger. How lovely that this philosophy is built right into the language.

## The Bio-Psycho-Socio-Spiritual Crisis

Not that you need reminding, but the summary below clarifies how much pressure a relationship endures when faced with infertility. If your relationship has been hit hard by the infertility, you need not feel inadequate. By looking at the summary of the multifaceted stressors, I want you to understand that if you're feeling wobbly, it's not that you are defective, but rather that you are lacking information. This chapter can fortify you for the battle.

Biologically:

- The diagnosis typically leaves women and men feeling physically defective.
- The intense mind and body experience of stress invariably kicks in.
- The medical setting is often intimidating—extremely intimidating for some.
- The prospect of surgery for women or men is frightening.
- The treatment usurps control of your menstrual cycle from nature.
- The treatment is demanding on body and mind.

Psychologically:

- Change is disorienting; unasked for change can be devastating.
- Self-esteem is at risk.
- Feelings of impatience and insecurity can be standard.
- Blaming the partner diagnosed as having "the problem" is a problem.
- Yet, thinking "I'd be pregnant if it weren't for him/her" is normal.
- Having "undiagnosed infertility" increases frustration even more.
- Loss of privacy is an adjustment.
- Loss of privacy of your private parts is an adjustment, for some, an ordeal.
- The takeover of your menstrual cycle leaves you feeling "hijacked."
- Mood swings caused by IVF drugs can escalate the temptation for conflict.
- It's most common to feel out of control and angry.
- Every twenty-eight days can represent another "death" if you don't conceive.
- Anxiety and depression go with the territory.
- Treatment is expensive; money worries can be very disruptive.
- Occupational realities cause even more pressure.
- For women, especially, there is an obsession with the calendar.
- The need to deal with insurance companies makes bedlam look appealing.

<u>Socially:</u>

- Feelings of isolation loom large.
- Babies seem to be everywhere, reminding you of what you don't have.
- Few know what to say to you; there is no social protocol.
- Resolving with whom to share this personal crisis is confusing and harrowing.
- You may have to shift the way you socialize with friends and family.

<u>Spiritually:</u>

- The question of meaning—why me—prevails.
- It is common but irrational to feel punished.
- A history of abortion or sexually transmitted disease may come to mind.
- Thoughts about connection/disconnection from a higher power can intrude.
- Moral issues loom large if you elect to "reduce" from a triplet or twin pregnancy.

## And Then There's Sex!

Sex can lose its luster with the pressure to time intercourse with ovulation. If you've been approaching the goal of pregnancy with diet, acupuncture, and mild ovulation-stimulation pharmaceuticals, you still need to have sex within the optimum range. Spontaneity flies out the window. This alone can affect a relationship negatively. Often, I've heard men say, "She only wants me when she's ovulating." Sex on demand has a talent for killing enjoyment, if not drive. If intra uterine insemination (IUI) or in vitro fertilization (IVF) is the appropriate treatment plan for you, baby making is removed from lovemaking altogether and, horror of horrors, it becomes unnecessary to have sex at all. While IUI and IVF are normal these days, they are not natural.

They say that power is an aphrodisiac. Infertility, to the contrary, by provoking feelings of powerlessness, can bring the flow of sexual juices to a screeching halt. No one signs up for this kind of sex life.

Try and picture this: Inject yourself with Follistim[12] and then hCG.[13] At the right time, put your legs in stirrups and try to relax while the doctor squirts hubby's ejaculate, collected in a sterile cup (pardon the pun), into your vagina with a syringe. It is as if the expectation of love making to produce your progeny has been put in a blender and someone pushed the liquefy button. Now what? Go home and enjoy lovemaking for its own sake, as a beautiful expression of your feelings for one another? Maybe. Maybe not.

You need to find a way to hold on to the feeling that love, not technology, is driving this quest. The tension of the emotional gridlocks that ensue from any aspect of this bio-psycho-socio-spiritual challenge, can be eased with some old fashioned TLC.

## Add Insult to Injury...

Your doctor tells you to have sex and then come into the office for a postcoital test. Good grief, lovemaking has now become coitus! The formality of medical science is starting to do something detrimental to the spontaneity and meaning of sex. You need to brace yourself. Here comes the judgment about whether your vaginal biochemistry is incompatible with your partner's sperm. How debilitating!

When it's suspected that there is a barrier between you and your fertility, sex can shift away from the heart and be reduced to an activity, a means to an end. Before you know it, your sex drive is driven to some distant planet. Are you too overwrought to respond to each other the way you used to? This goes with the territory (more likely if sexual abuse is in your history). Even if sex has been consistently satisfying, worrying about your prospects for pregnancy could leave little room for enjoyment. Feelings of desperation can trump the heart's involvement in your sexual connection to each other.

It's no picnic for men either. Sex on demand—"I'm ovulating, honey! Get over here!"—is a turnoff, and turns a choice into a constraint. How unsexy. It is par for the course for a man's "machinery" to shut down from time to time. With in vitro fertilization, hubby's ejaculate makes its way to your uterus by way of an intermediary. I have never known a man to be excited—pardon another pun—to ejaculate into a cup. One of my patients spoke with disdain about having to produce a "specimen" in what he called "the Boom-Boom Room," a closet-sized space replete with a few *Playboy* magazines.

Neither you nor your husband signed up for sex to be irrelevant, did you? This is the world of in vitro fertilization. You should acknowledge the pressure that infertility puts on your sex life. And you should know that, among your peers who are out of the mainstream like you, it is normal for libido to take a serious nosedive and for erectile dysfunction to be an understandable symptom of this stress. Female sexual responsiveness is just as vulnerable. If you can accept this, then at least you have one less excuse to fault yourself or your partner for faltering lovemaking.

> At the beginning, the infertility did cause conflict. It was very stressful before we actually got a diagnosis, and we were just trying and failing. We were in the honeymoon phase of our marriage, which came to an abrupt end—sort of a screeching halt. Then we found out it was male factor. He didn't want to talk about it, and it was all I wanted to talk about. We actually went to a couple's counselor. Once we realized that his guilt explained his resistance, but his resistance to talk was giving me incredible anxiety, we were able to have a lot more sympathy for each other, which brought us to a more mature stage of our marriage. —*Carol - writer, New York, NY*

## What's Love Got to Do with It?

Couples form relationships based on a resonance that is experienced as love. But underneath conscious awareness, there is another resonance, which explains why any two people match up. It is common for the life stories of partners to have similar or opposite emotional responses to the same issues.

There's a wild card in the deck. Life stories are the "programs" that get imprinted in early childhood when we're minding our own business, piling one alphabet block on another. These programs lodge in the unconscious mind, and they influence how we cope.

Our unconscious mind is the wild card, and it does its very best to form a worldview that is designed to keep us safe, even if that worldview is illogical. In some homes, the way people interact may be entirely dysfunctional, but our young minds do the best they can in formulating a plan for coping.

Coping is difficult without adequate and accurate information. Imagine trying to write a book if you only have fifteen letters of the alphabet and none of them are vowels. We're all handicapped in some way—we all have a thing or two to learn about coping when the going gets rough. It's not until we're faced with a challenge such as infertility that we realize that the manifestations of our belief systems, which made it possible for us to get through life so far, may now be making us, and the important people in our lives, uncomfortable.

What Brenda, a writer, referred to as personality problems, I'd prefer to call personality traits. As I mentioned in the last chapter, the common personality traits that I focus on later in this book are anger, negativity, insecurity, impatience, fear, and lowered self-esteem.

These normal characteristics can become personally or interpersonally troublesome when pressures from infertility exaggerate them. What this means is that the love you feel for one another is free to flower and grow until the weeds of infertility threaten to strangle it to death.

**"So if I broke your heart last night, it's because I love you most of all."**
—*Mills Brothers,* "You Always Hurt the One You Love"

Early environments define what love is. Unfortunately, the definition of love is often remote from its real meaning. In some relationships, love means suffocation. Love is sometimes conditional and sometimes a business deal. It can have expectations of dependency or one-upmanship, and there are other distortions. Worst of all, believe it or not, love can have its basis in what more closely resembles hate. The kind of love you want is love that's ruled by the heart, but it takes work and self-awareness to get past dysfunctional coping styles that your role models may have demonstrated.

Our emotional inheritance has such a powerful influence that it may as well be genetic. We learn what we live, and we can't escape from our emotional inheritance anymore than we can escape from having blue or brown eyes—unless we *become aware of* the distortion and put effort into learning new ways. To add insult to injury, distortions about how to love can be out of conscious awareness. This heritage doesn't give us the best chance at love since the reality is that we end up bringing scars or wounds into our relationships. The natural inclination is to find someone who will not disturb our wounds—lots of luck with that! In the throes of an infertility saga, the chances of a wound remaining unprovoked are slim to none.

The healthiest relationships are the ones where a couple works out its wounds *with* each other, not *on* each other. It is unrealistic to expect that, wittingly or unwittingly, your partner will never ruffle certain feathers. "With each other" means that you and your partner are each willing to look at why *you* feel that *your* feathers are ruffled. This means looking in the mirror, seeking self-awareness. "On each other" means that you and/or your partner hold a rigid adherence to the notion that the other person will *never* say or do anything that opens an old wound and then have a fit when this "rule" is violated.

Of course, it's impossible to restrain indelicate remarks 100 percent of the time, especially if those remarks spring from the unconscious. But, know that when a wound is provoked, as is predictable, anger ensues. Each wants the other to change. Without self-awareness and other awareness, this is a no-win trap. If only he/she did not [fill in the blank] you would have the best relationship! Sound familiar?

Under duress, the aspect of your connection to one another, which is based at least in part on each person's emotional wounds, is bound to zoom out of hiding. To say it another way, the aspect of your relationship that was written in invisible ink is now being held up to ultraviolet light, and the instructions are in bold print.

This painful process is unavoidable, but it's also an *opportunity to heal* individual and relational problems simultaneously. It can be a blessing in disguise, if you let it. "Good grief," you might be saying, "not more of that 'crisis as opportunity' crap!" Actually, yes, that's exactly the point, because issues between couples are usually so well matched that opportunities to heal wounds are greatly enhanced if both people are dedicated to understanding what needs to change in *themselves*, not the other person. You will see later in this chapter, with the case of Amelia and Angelo, how the gridlock that couples get into can actually serve as a doorway to the most satisfying and relieving growth.

It takes openness, courage, determination, and patience, as well as love, to face our wounds and heal them. Emotional healing is not an easy, one-shot realization, just like a surgical wound doesn't heal overnight. It is a process of identifying, articulating, negotiating—and inevitably messing things up many times—before it is possible to arrive on the other side of the mountain. As difficult as this process is, it builds true intimacy and deepens love, so it's worth every step you have to take.

## Commune – icate

You've undoubtedly heard it before. The most important key to building intimacy is effective communication. When communication skills are sufficiently developed, the road will still have some bumps, but you would have the shock absorbers you need to help smooth the ride.

On this journey, the ability to expose and express vulnerabilities is a requirement. This presents two problems. First, if early life experiences mandated that self-awareness *must not* develop, then how can you expose what is really true for you if you don't know it yourself? Second, our egos have an aversion to revealing vulnerabilities because if there needs to be any compromising, they worry that they might lose their top dog status. Egos do not understand that they can get promoted to an even better job. And keep in mind that the entire infertility experience beats you to a pulp, so vulnerabilities are set to fire off like a jack-in-the-box even if you have no idea what's happening.

Another requirement in growing your love affair is the determination and flexibility to make necessary changes. Personalities cannot help but get set in stone into a default mode. Therefore, you may not have the flexibility needed to be self-aware and other aware if that was not cultivated in your early years. Furthermore, the ego that embedded itself may not understand the patience, determination, and perseverance needed to tolerate the disruption of change—even when the change is sought after. One common setback on what starts out as a successful path to change is for the ego to flop back to business as usual—its default mode.

When a default mode creates dysfunction, *it is important to remember that our default mode is not our fault.* We take on characteristics of what we "marinated in" as youngsters. However, we can choose to take responsibility for changing the default mode that developed. In the process, we can heal a rigid stance and enhance our relationship with ourselves and others.

To heal is to remember inner resources by way of letting-go coping. We all have these resources, but they may have either gone into hiding or withered from disuse. Mind/body awareness is the main ingredient for healing. Letting-go techniques like hypnosis or self-hypnosis can help to shine a light on your inner treasures. Once you discover, or rediscover, your resources, it becomes possible to use them to follow through with restraining old behaviors and installing new ones. The budding science of neuroplasticity has taught us that if we change our minds, we can actually change our brains.[14]

Throughout these chapters, this book focuses on self-awareness. The hazards get trickier when you not only have to understand yourself, but also your partner, when you never really learned the right tools to do this. Amelia and Angelo's story illustrates that even when resources are undeveloped, repair is entirely possible.

## Default or De Fault of Amelia and Angelo?

Angelo and Amelia, a couple who both work in theater, were each on autopilot with their respective emotional inheritance; each had an emotional inheritance that was destined to contaminate and pollute their love. From the outset, their real love for each other was doomed to lead to distorted expectations. And the infertility struggle just upped the ante.

In Amelia and Angelo's first therapy session, they admitted that their "relationship was volatile before" they started trying to have a baby, but that the fertility quest "made their differences stronger." By the time they landed in my office, they had already been going through the fertility struggle for two years. They had endured four IVFs and had conceived on the third but suffered an eight-week miscarriage. Angelo was the one diagnosed with the fertility problem and Amelia, now thirty-eight, had begun to feel that her age might be a factor, too, even though her FSH[15] levels were consistently excellent. Both were overwhelmed from all of this and felt squeezed by the cost of pursuing treatment.

Some people don't understand themselves *or* their partner. This was not the case for Amelia and Angelo. They knew why their behavior toward each other was provoking more and more anger and frustration. But, knowing there is a problem, and knowing how to *solve* the problem, are two different things entirely. Even if you know you need to change, you also need to be committed to doing the hard work to make that change happen. Paying lip service is not enough.

Both self-awareness and understanding what your partner is going through (other awareness) is important. If you can be honest with yourself about how you behave, you can understand how your behavior affects the other person. With this honest appraisal, you are in position to learn the way your partner's unwanted behavior might then bounce back to you *in response to your behavior*. Your partner needs to be honest as well. This is sophisticated business and not for the lily-livered. Fortunately, Amelia and Angelo proved to have the stamina, the moxie, and, yes, the love to put self- and other-awareness to the test by creating necessary change in their own behavior.

Despite the fact that each of them admitted to what they contributed to the emotional frenzy, neither had a clue as to how to stop their actions. Amelia knew, for instance, that when she is angry she "takes it out on him and gets demanding." Angelo knew that he is "not a good communicator" and felt that he needed "anger-management training."

In their first session, this couple demonstrated the depth of their gridlock by putting on a display of ferocious anger. Frustrations and accusations flew fast and furious until Angelo got up and stormed out of the room. Your relationship may not be anywhere near this volatile, but it gives you an idea of how heated things can get if you lack the information that you need to cope effectively with your differences.

## Shifting into a No-Fault State of Mind

Amelia's definition of love formed in a home where her parents reassured each other of their connection with words—vicious words. Amelia was "recruited" into the role of the mediator, with the unspoken agreement that nothing would ever really be mediated. For Amelia, there was an illusion of having the power to mediate. The illusion was really a delusion, which locked Amelia into a belief that she was close with her parents. In her family, discord was the family glue. Hurtful words flew around like shrapnel, and in the end, had no meaning, except to keep mom, dad, and daughter on the battlefield that the parents established to protect themselves from the "treachery" of intimacy. Yes, folks, some people will go to great lengths to avoid behaving lovingly. Some people arrive in adulthood thinking of tenderness as weakness.

Amelia had been skipping off into the sunset thinking she was close with her parents because she always said what she felt. She never put together that her words of wisdom—and they were—did nothing to alter the parents' verbal assaults on each other or on her. The satisfaction she never got from her parents she wanted and demanded from Angelo. Unconsciously, she played out with Angelo her unmet longing for true communication. To Amelia's way of thinking, when she imitated the verbal abusiveness of her parents, she presumed she was communicating with her husband because that's all she knew and that's all she was ever taught. In reality, her behavior only prompted the dismissal of her feelings by Angelo because her style of communication fostered a self-protective stance in him. The power of unconscious processes and the perfect match of their dynamics created an exact replica of her family's "communication."

Angelo, in the meantime, inherited his family's pattern of discord. Whenever there had been an emotional issue, Angelo's father's discomfort with feelings mandated that he either remained silent or made a joke. Both responses drove his mother to shrewish behavior, which of course, justified to the father's thinking that he was right to remain silent and disengaged. Whereas discord was the glue in Amelia's family, the pattern of discord in Angelo's family resulted in divorce. So, when things got particularly out of hand for Amelia and Angelo, the D word, as they called it, ricocheted off the walls of their home.

Amelia had always found Angelo's inability to communicate in the way she needed him to as outrageous. She said, "When he shuts down, it sends me into an absolutely crazy place." This, of course reminded Angelo of his mother, and what he grew up watching his parents do, so he would automatically shut down like his father.

Furthermore, Amelia would talk about how infuriating it felt when she would need to discuss something with Angelo and he would say, "I don't know how to do that. You have to help me." She said, "[I] thought it was the most ridiculous thing in the world that I would have to teach him how to be in a relationship." Truth be told, he did need guidance, but she completely missed that he *wanted* to communicate with her. Her ferociousness froze his capacity to find his thoughts and words until his only release was to explode in anger. Anger-management training would have served no purpose in a relationship where Amelia's contribution to the vicious circle would have been left in place.

What was below her conscious awareness was that her inclination to pick fights replicated her family's definition of love, in which words created the illusion of communication but were meaningless. Amelia took responsibility for "chasing him down the street to finish a fight." She said, "I had to win." What she was actually doing was chasing him down the street as a way of finishing "loving" in her family's tradition.

Below Angelo's conscious awareness was his procrastination about getting his fertility problem treated. Consciously, he was aware of going to great lengths to avoid conflict, but unconsciously, if he did not do his part by getting evaluated for possible treatment of his sperm issue, he could maintain the familiarity of living with a raging woman while circumventing his guilt because of his low sperm count and his fear of failure. He continued his family's tradition of hostile silence or joking because that's all he knew how to do.

You can see the way that Amelia and Angelo's dynamics fit together perfectly and, therefore, presented the opportunity to transform working their problems out *on* each other to working out their problems *with* each other. Happily, they were able to do this. The deepening of this couple's self- and other-awareness, and their willingness to make changes in their own behavior, shifted their perspective from feeling, as they put it, "effed up," to the realization that they were each at the mercy of their emotional inheritances, which was nobody's fault. Their evolution is inspirational and possible for other couples willing to do the work.

## Interventions That Supported Their Evolution

Angelo and Amelia benefited enormously from their willingness to take guidance. Perhaps the five interventions that we worked on in their sessions will benefit you as well:

## 1. Learn and practice a calmer body state.

With Amelia and Angelo, it was beneficial to use the letting-go technique of hypnosis to help them break through old beliefs and attitude and move forward together. They knew they wanted to love each other in a different way, a way that was a departure from all that they had known. Hypnosis will be discusses in detail and self-hypnosis will be taught in chapter 7, but for now, suffice it to say that they learned to let go of the body tension that maintained the default mode and enter a body state more conducive to the kind of love they wanted. From this calm vantage point, they could see the logic of their entrenched behavior and the foolishness of it.

They each understood that their behaviors had become automatic. They wanted to replace them with actions that accurately represented their feelings. Their compassion for each other grew as they watched each other struggle to change. They learned that in spite of the strong gravitational force trying to pull them back to their old behaviors, they could succeed if they continued to work hard at their dual goal of growing a love affair as they grew a space into which they could bring a child. First and foremost, it was important to *get comfortable with getting comfortable.*

Eventually, they started to enjoy their capacity to be a team. Amelia softened and became more compassionate as she understood Angelo's struggle with his low sperm count. Angelo responded and followed up on his treatment. Ultimately, they worked through their choice to used donor sperm. As a back-up plan, they began to gather information about adoption. They helped each other adjust to the swirl of emotions that cascaded from this shift in thinking, including Amelia's sadness about not using Angelo's sperm. They developed the flexible thinking that they needed to navigate their vision quest.

## 2. Learn and practice simple acts of kindness.

As the tension between them continued to simmer down, Amelia and Angelo were reminded to demonstrate simple acts of kindness. They had lost touch with the pleasure of hugging each other just for the sake of it, with kissing each other hello and good-bye, with expressing appreciation while simultaneously making eye contact, or with simply reaching out to touch a shoulder or hold a hand at times of sadness or disappointment. At first, it was scary for them to try these simple aspects of loving behavior because their egos contained the imprint of a different agenda. These simple acts of kindness became "homework" assignments so they could adapt to new behaviors and avoid slipping back into their old patterns.

## 3. Build effective communication skills.

Angelo needed to find his voice and Amelia needed to hold hers back. The sessions were a safe space for each of them. Finally, Angelo was able to speak about his "difficulty in expressing himself emotionally." During the sessions, it was critical that Amelia not interrupt Angelo. It took guidance, but they both cooperated because they both wanted to change. Both learned to tolerate the discomfort of silence until Angelo could find and express the words that accurately represented what he felt.

Along the way, Angelo was slowly able to realize that Amelia was beginning to understand what he was doing when he would make a joke and could appreciate how he was putting in the effort to change that behavior. Angelo said he was learning to listen and was pleased to discover that they could have a conversation. Even better, he was able to take initiative, identify misunderstandings, and fix problems before they blew up. Amelia really appreciated his newfound confidence, and she claimed victory as well because she had begun to resist her old pattern of firing off.

## 4. Understand nonverbal communication.

This couple needed to tune in to the subtlety of nonverbal communication. One session was particularly important in facilitating this breakthrough. In anticipation of the third donor insemination, I asked Amelia to anticipate what she would need from Angelo if the ovulation-stimulating drugs overpowered her emotionally. She said, "If my body reacts badly, and I'm hurting, in pain and 'angsty,' I could pick fights. When I get stressed, my discomfort level goes from zero to ten in like three seconds, and I start lashing out. I know it's not a good thing, and I need to stop doing that...but he immediately engages in my frenzy, and what I need is for him to maybe step aside and say to me 'this is happening because you're on these drugs... everything's gonna be okay.' (She paused.) I guess that's the thing. I think I just need someone to keep telling me that everything's going to be okay because it feels like it's not."

This was great. But, Amelia was still having problems with letting Angelo hold her and express intimacy, so how could he reach out and connect with her? Her body language was pushing him away. I pointed this out to them and suggested that Angelo reach out to her physically and that she let him. In this session, they learned that physical contact, even in silence, can be loving and comfortable and should be part of their repertoire. This session began a new phase in which they could experience a silent embrace as a loving gesture. They commented that they were thrilled that this new pattern, representing an accurate expression of love, would be the environment into which a future son or daughter could be brought.

### 5. Do not lose sight of the love that goes into making a baby, even by hi-tech methods.

We talked about how important it is to humanize the technology, so you don't lose sight of love in all the technical stuff. Amelia and Angelo experienced the letting-go coping method of hypnosis simultaneously and it helped them connect with each other. The script from their hypnotic interlude will be included later in chapter 7.

## Stressed is Desserts Spelled Backward[16]

Amelia and Angelo earned their just desserts. Their hard work has allowed them to love and respect each other in new ways. They now understand that the toxicity of what they each earned in their early environments fit together perfectly. Of course, what they learned in those early years sank underground and became their unconscious beliefs. But, now they have mastered stepping back from what had been their automatic patterns, an achievement that was like stopping a runaway locomotive. They understand that the glitches that they still get into are the residue of where they used to be. Individual histories follow all of us, but we can learn to let go of what doesn't serve us well. Due to their hard work, the glitches are fewer and less intense. They catch themselves and replace where they *start* to go with where they *want* to go.

When Amelia and Angelo were asked what was going through their minds when they were first trying to conceive and fighting like cat and dog, their answer wasn't surprising. With their greatly improved communication skills, they were able to say that deep down, as much as they longed for and wanted a child, it upset them to think about bringing a baby into a hellish environment. By this point, the couple was well on their way through the adoption preliminaries and had made some difficult decisions together. Angelo was taking more initiative in the process, and Amelia could relax now, feeling more supported.

This couple put enormous effort into changing the environment into which they would bring a baby. Their ability to function as a team grew beautifully, and they felt inspired to reverse an earlier choice and decided to give their last vial of donor sperm a go. This time it worked, and Amelia and Angelo are now proud parents of a little boy.

## Vive la Différence?

If you were to ask women who are in the best of relationships about the most common issue that emerges during a fertility crisis, many of them would say, "He doesn't feel things the way that I do." This can be a source of disconnect, even in the relationships that are tightest. The best-case scenario happened for Jenn, a singer/songwriter. Her growth got her to a place where she realized, "If Josh doesn't understand me, it just means that he doesn't understand me right now."

To a certain extent, this is the difference between the way that men and women approach problem solving. Men tend to *think* about how they can fix what's broken—mostly the car or the plumbing. Women tend to *feel* how the problem is affecting them. Long live the difference when the man is more clear-headed and can provide the voice of reason to solve a problem. As Amelia put it, "tell me that everything is going to be okay."

But because this experience is not about a car or a toilet, women, who want to solve the problem as much as men, typically prefer that la différence be obliterated. Women automatically want men to just get it—post haste. Women dealing with infertility often say, "I'm angry that he is not as angry as I am." Infertility is boot camp for men who "come from Mars"—and the women from "Venus" aren't very patient when it comes to tolerating the confusion some men feel as they try to enter the world of emotions.

Furthermore, it takes getting banged around a bit before most men can say, as Brian said to Stacie, "What happens to you happens to me." Brian eventually became upset when no one asked *him* how *he* was feeling. Getting comfortable with feelings, or at least tolerating the discomfort of feelings if you're not used to them, often happens when men realize that it makes sense to shift to a place where, for example, they can grieve a miscarriage, too. It is a victory for a man to understand that a failed IVF is the son or daughter that he, too, will not have. These realizations are what bring a couple closer.

It is understandable for a man to feel hesitant to approach an overwrought wife if he is uncomfortable with his own emotions or unpracticed in supporting someone else's. Sometimes men, when they were little boys, observed women in their families, most likely their mothers, who were emotionally irrational. If their father literally or figuratively abandoned their mother, as was the case in Angelo's family, then chances are that the little boy was somehow expected to provide his mother with support—a role he would be too young and ill-prepared to fill. If, in his adult life, this man's wife is in a highly charged emotional state, he would feel tossed back into his boyhood where he felt helpless. In this case, a man who runs from emotional upheaval is really running from his mother. If couples understand this problem, then they can work on it.

In fertility cases where the diagnosis is male factor, women are often overprotective of the man's feelings. There seems to be a tacit agreement that sperm issues are really manhood issues rather than the malfunctioning of a body part. Real growth mandates letting go of this socially supported, macho attitude, and facing reality for what it is, namely, the manifestation of a mind/body interplay that he, and they, can work through.

Carol expressed the other side of the coin in the male fertility department. She said, "I know my husband would not be ashamed to tell people why they weren't conceiving. It's that *I* don't like to tell people, 'Oh it wasn't me. It was my husband.'" She expressed an exquisite sensitivity that was about being a team. "For us, the biggest change was in the sympathy we developed for one another, the confidence in the strength of our marriage, and the feeling that there wasn't anything that Tom and I couldn't handle together."

While this generosity of spirit is fine and dandy, do not berate yourself if you go through a period of resentment if the male partner is the one with the diagnosis. It is normal to feel upset that you would be pregnant if he were fertile. How could you not be upset? The real issue is learning that, sometimes, the best we can do is just state how we feel and know that the other person is not taking it personally and can validate our feelings. The same would be true if the woman is the one who has the diagnosis. Sensitivity toward one another is always a good thing whether for male, female, or undiagnosed infertility.

Nevertheless, bringing feelings into the light of day and getting comfortable with them is an important skill that you can learn. With open communication as part of your repertoire, relationships become infinitely stronger than they were before. With adequate empathy and communication, both can feel supported, and it can come to be okay that they may not be experiencing the problem similarly.

Let's define the kind of communication that can make the difference. Remember, Amelia would chase Angelo down the street to "finish communicating." She was looking to feel understood. But *good communication is as much about understanding where the other person is coming from—what feels really important to him or her—as it is about expressing your take on things.*

Infertility is hugely complicated. When it comes to effective communication, here's the takeaway: Sometimes listening to your partner needs to trump speaking. Why? Because the real solution to infertility is bringing home a baby. For now, the only guarantee you have is that you will find yourself in situations that are bound to be fraught with frustration, impatience, negative expectations, and suffering. It's best if you can find a way to value hearing as well as being heard because that might be as good as it gets. The exercise at the end of this chapter will show you how to do just that.

## Meanwhile, Keep the Love Flower Blooming...

Love is about openness, expansiveness, abundance, and receptivity. Love is fuel for joyous living. Love takes us inside and outside of ourselves. Love is the place where any two people can meet softly in kindness and understanding. Love is the underpinning of all that is good in this world. Love is what makes us want to reproduce.

The opposite of love is fear not hate. Anyone raised in a fear-based environment could come into the infertility experience in a fear-based body. When you put this together with how much fear cascades from the infertility process, you can appreciate the obstacles to keeping love vital. One patient, Cecile, put it so perfectly. She said she felt the infertility experience was a test of her marriage that would prepare them for the tests of parenthood. Voila! Now you know why one message of this book is the potential adversity has for growth.

## Exercise One

The following exercise is an adaptation of the work of Harville Hendrix, delineated in his book *Getting the Love You Want.*[17] As you learn and practice this technique, you will come to see the merit in no-fault communication. The idea is to defuse tense situations not to solve what for now may feel unsolvable. The punch can be taken out of deadlocked issues that feel beyond resolution. When there is no easy resolution of any situation particular to infertility, each person gets to have his or her say and feel understood. Each person gets to appreciate the mindset of the other. When discord can be simmered down, you can appreciate that for now—that's as good as it gets.

In this exercise:

- The SPEAKER talks for several minutes about an upsetting situation.
- The LISTENER *may not interrupt* and must listen attentively.
- When the SPEAKER stops talking, the LISTENER repeats what the SPEAKER has said and then asks, "Is that right?"
- If it isn't correct or if part of it was incorrect, the SPEAKER repeats what the LISTENER missed.
- When the LISTENER correctly repeats what was missed, or if it had been stated correctly the first time, the LISTENER says, "Is there more?"
- If there is more, the SPEAKER continues, and the above process goes on until the SPEAKER truly has nothing else to say, and the LISTENER has repeated everything correctly. It is important to "empty" the emotionality out of the issue by saying everything that is bothersome, even if there is nothing that can be done to change the situation.
- Then the couple switches roles, so that the SPEAKER becomes the LISTENER, and the LISTENER becomes the SPEAKER.

This style of communication should be repeated as often as is needed. You will see that boiling issues can be simmered down, especially once couples learn the value of not interrupting.

## Exercises Two, Three, and Four

The first two exercises below are nonverbal, although consider it "legal" to burst out laughing.

**Exercise Two:** Spend several minutes coordinating your breathing.

**Exercise Three:** Spend several minutes coordinating your blinking.

Are you catching on that in order to do the above exercises, each of you needs to be aware of the other? This next exercise is nonverbal as well and it will teach you to tolerate silence.

**Exercise Four:** Lie down next to one another. Give yourselves time to settle in. Feel each other's presence. Think positive, loving thoughts about the other. After a while, hold hands or put your arms around one another and continue holding positive, loving thoughts. This is about feeling your connection to one another.

This is not as easy as it sounds. There are likely to be intrusive thoughts that you may have to master, perhaps with a few minutes of the relaxation response or focused breathing. You may feel your or your partner's discomfort as well as comfort, since we tend to want to fill silence with words or actions. You cannot do this exercise incorrectly. The goal is to become aware (there's that word again) of the power of nonverbal communication.

Chapter 4

# Infertility's Impact on Friendships/Family: From Involvement to Isolation

Presto-chango! You're out of the mainstream. Slightly more than four of every five couples in their childbearing years are continuing on their path toward parenthood. You're left in the lurch as part of the growing minority, but a minority nevertheless, that needs to bring baby making out of the bedroom. You need help. It isn't fair.

Especially when you're going through a crisis, isolation can make your pain much worse. Feeling connected to a support system is a comforting remedy for most people, yet the infertility crisis presents a dilemma. If you reach out to friends and family for solace at this difficult time, you relinquish the privacy you expect around procreation, or you isolate with your secret. What should you do? Feeling torn between wanting the privacy that everyone else has around making a baby and wanting relief from the pain of isolation may leave you feeling like you have two heads, each with a mind of its own.

In this chapter, you will see the range of responses from isolation to overexposure that should help you to identify what resonates for you.

## Bye, Bye Privacy...Nice Knowin' Ya

If you want to get pregnant, you can't stay in the sanctuary of your bedroom. Your sex life and your menstrual cycle used to be nobody's business but yours. Why do your private parts need to be under the glare of fluorescent lights? When I interviewed Donna, a children's book editor, she said, "My doctor saw more of my vagina than my husband." A working mom of two, Donna received bleak odds of conceiving. She beat the odds because, among other things, she was willing to put her legs in stirrups.

You're no longer in the life that you knew, and you may not yet have settled into the new reality. Welcome to no man's land. Meanwhile, your ovaries, uterus, and fallopian tubes have become the focus of scrutiny. This feels dehumanizing, especially because your broken heart is left out of the story. You remind yourself that the exquisite advances in reproductive medicine are inspirational. Then you get your period again. You're bummed out. Where do you turn for comfort?

## To Tell or Not to Tell...That is the Question

It's one thing that you now have doctors in the middle of your relationship. Should others be there as well? This is a gigantic issue. Are you familiar with John Donne's poem "For Whom the Bell Tolls"? His words "no man is an island" are a truism. We need each other. We're hard-wired for connection. Even if the most important connection—the one between you and your partner—is graced with solid, mutual support, can the relationship tolerate the many pressures of infertility without a safety valve? Can any two people be *everything* to each other at all times? Does the containment of this information make you feel like you are about to explode or implode? Would venting your feelings to a select person or people be an appropriate outlet for you? Are there people out there who could actually support you in ways that would help?

Deciding if, or whom you need for support can take on an agonizing dissonance. A battle within yourself and/or a difference of opinion between you and your partner can feel like the tumult a washing machine makes when all of the towels land on one side of the rotor.

Recently, a couple I'll call Elizabeth and Peter, came to my office in a state of discord about keeping their circumstances private. Elizabeth was suffering greatly and wanted their situation kept between the two of them. In a twist of what is usual, Peter felt compelled to open up to his family about their reproductive status—without consulting her. More often, it is the wife who is overly close with her mother, in which case the husband can feel like little more than a sperm donor. In this family, Peter was the one who behaved as if his sister and his parents were more important than his wife.

Not surprisingly, there was bad blood between Peter's family and Elizabeth. He was unable to understand what the big deal was in exposing their situation, virtually eliminating the chance for a meeting of the minds with his wife. Resolving the privacy issue is one place of many where relationships are at risk to destabilize.

This couple was at risk. They were invested in the blame game, not in self-awareness. The discord between them was leading the parade, and they were deadlocked on the privacy issue, especially because the horse was already out of the barn. They were in the early stages of the diagnosis and treatment, and their level of emotional reactivity was sky high. They were resistant to taking guidance and resistant to cutting each other slack and moving on. Perhaps this couple will expand into self-awareness and awareness of each other's feelings, so that mutual compassion and effective communication can evolve. Perhaps not.

There are multiple quandaries involved in choosing to go public rather than maintain privacy. What do you do if your boss doesn't know what you're going through? How do you show up at work late, leave during the day, or not show up at all, given the frequent medical tests and procedures? How do you or your spouse travel for business if the trip coincides with your fertile time?

Your personal life has its own complications. What do you say to your parents if they start hinting about wanting to be grandparents without realizing that they are adding to your pain? What do you do with your feelings when the pregnant bellies of your friends, relatives, and neighbors seem to surround you? It's an ordeal to maintain a poker face when wherever you turn, you see people with babies.

A couple who is on the same page and experienced in demonstrating compassion toward each other has a shot at swimming upstream to get back into the mainstream without anyone knowing besides them and the doctors. Much depends on your capacity to both contain and discharge the pressure.

How do you decide which friends or family members will be appropriately supportive? If you personally or you and your partner as a couple are at an impasse, getting professional help can smooth the way.

## Whom Do You Trust?

In the past, have you and your spouse been casual and comfortable about letting others into your private space? Do you had a history of trusting people and, more important, have the people whom you have trusted earned that trust?

If you have certain friends or family members who are blabbermouths, it would be logical to withhold information from them. Perhaps you know someone who would feel like a big shot if they had "juicy news" about you to publicize. These people tend to be unable, truly unable, to respect your request not to tell anyone else. These are the kind of people who do not belong in your inner circle.

It's not always possible to know who will come through for you. You might need some resiliency if whom you choose to open up to disappoints you, even if they keep your secret. They may not know what to do with the information in order to give you the kind of support that you need. Or, maybe you've exposed your news to a trustworthy person of valiant intentions who says something that he/she thinks will help but offends you. Sometimes trustworthy people inflict wounds unintentionally. Resiliency is important here as well. Communicating your hurt so that you have a chance to repair the damage is important, too. The infertility nightmare can be a slippery slope, even for the people who love you and want nothing but the best for you.

It might not have occurred to you that you can share aspects of your struggle without revealing all of the details. For instance, if you feel that you would be negatively judged for using donor egg or sperm, you may want to keep that aspect of the situation under your hat, even as you share that you're going through an IVF. If you want to make sure that *you* are the one who tells a future child about his sperm- or ovum-donated origins, then holding this information close to the vest makes sense. Yet, select people can still provide a safety valve where you can release some of the intensity of the struggle.

## Stop the World...I Want to Get Off

Trusting others with the intimate details of your life is not the only consideration. You have a right to protect yourself from exposure to pregnant people and babies while you are so raw. What do you do when holidays or other celebrations appear on your calendar? What do you do if friends, families or co-workers expect you to be at an event, but you fear that the pain would be too great for you if you went?

Furthermore, there can be a double jeopardy at some family gatherings: an expectation of a celebratory mood and a pattern of family conflict. The expectation of gaiety when you feel anything but gay and the anticipation of frenzy in the bargain can generate enormous distress. Now add those "innocent" questions—What's wrong or Why aren't you smiling or Why don't you two have children yet? Maybe you feel obliged to take a forced gaiety stance. These situations can produce emotional indigestion. What's best for you? Not to eat the "offending food," or to hope that you'll find an effective "antacid."

If you normally enjoy your family, you still need to give yourself permission to bow out of an event. The chaos caused by family dysfunction might not be an issue, but the presence of babies and bellies might be. What do you need to do about disclosure in this case? Can you go off to the Caribbean at Thanksgiving if no one knows why?

The issue of privacy looms large when faced with the inevitable invitation to a baby showe, or when someone you know gives birth. These instances rub salt in the wound. If you do appear at a baby-related event, it could be harrowing if no one knows what you are going through. You're stuck with acting differently than you feel. Some can pull that off. Can you?

> Joe and I went into this thing thinking that we had pretty great and perfect families. And then once the infertility hit hard, all of a sudden they started behaving differently...doing things that were all too human, but they weren't the responses that we wanted. They were hurting us, though not on purpose. But, the process of going into a therapy environment and formally dealing with these things made a big difference to us. Had we not done that...I don't think we would have worked through a lot of this to the clear-headed level that we did nor would we have gotten the benefit of "knock-on skills" for the future. We have the confidence to assert what is okay and not okay when it comes to keeping them in line as grandparents. — *Nanci, banker, Dallas, Texas*

## Self-Awareness Can Be So Freeing

It is your birthright to make the choice of maintaining full or partial privacy or scrutinizing what kind of help you need and who could best give it to you. What about deciding if you can tolerate being at your sister-in-law's baby shower? It can make a big difference if you tune in to your feelings and your partner's. You and your partner are best off if you have the information that you need to make tough decisions as a team. The ability to communicate would clearly be a plus. This is harder to do if you do not know yourself well. Organizing a list of pros and cons will clarify your thinking to a certain extent.

The capacity for assertiveness is another important quality. You need assertiveness to establish appropriate boundaries. If you find yourself short on these skills, let me remind you again that getting professional help can be relieving and productive.

If you are clear and without conflict about keeping the information under wraps, good for you. Usually people who don't reach out to others have very good reasons for it. There are many variations on this theme. Do you connect with any of the following stories?

## I Love You, Honey, But...

Feeling isolated as a couple is one thing. Feeling estranged from each other is something else and should be given immediate attention. But sometimes women who are in great relationships feel the need for other women. Monique, a lawyer/MBA/entrepreneur and now a stay-at-home mom, said that for her, a certain kind of aloneness developed. Even though she and her husband were tight teammates, she made the distinction that it was *her* body that was at issue. It was *her* sense of womanhood that felt damaged, and she needed to find other women who could be inside the experience with her. She said, "It was like [I was in] an underground movement to find a female friend who would know what I was feeling. And then I worried about the possibility of loss of that friendship if one of us became pregnant."

Monique found compatriots despite her worry that the solution could create another problem. She was right; friendships can shift like snow in a blizzard. But so much depends on the emotional makeup of the people involved. Some friendships remain predictably satisfying until the going gets tough. If a buried wound for one or the other becomes exposed, the solidity of the friendship might be shaken.

Friendships that weaken or wane are one end of the spectrum. Cecile, a retired lawyer and aspiring writer, was pleased to report her experience on the other end. Her anguish about the strength of a friendship proved to be unnecessary. She took the risk of asking her best friend, who was struggling with infertility herself, to be godmother to her son. The offer was enthusiastically accepted. Cecile's friend was able to maintain an awesome open-heartedness. We can speculate that her friend was committed to becoming a mommy...whatever it took. Far-sighted belief in a positive outcome would mitigate understandable jealousy. Friendships can thus thrive.

By the way, men could very well need to reach out to other men, but social norms being what they are, there is not much talk of this.

## Uh-Oh...

At first, neither Shari, an event planner, nor Steve, a college professor had wanted children. But when the role of "bachelor couple" played itself out for Shari, she was suddenly struck with an awareness of wanting a child. Her husband, though surprised at first, came around to agreeing.

When the doctors assured her that she would conceive with inseminations, Shari felt it was no big deal to go to the appointments alone. As her fertility quest became complicated and dragged on, her suffering led her to my mind/body support group.

Membership in the mind/body group brought with it an expansion of self-awareness and slowly a realization emerged that she felt alone in her marriage. She had worried that Steve's agreement to have a child was fragile, which it was not. She said that she did not want to push him to go to doctor appointments with her because she thought "it would piss him off." Her developing awareness of a need to communicate, and a growing sense of entitlement to speaking her mind, enabled her to share her feelings with Steve. He easily opened with empathy to the aloneness that was plaguing her. He came with her to medical appointments after that, and they resolved the fertility crisis together when they adopted their daughter. The bond in the marriage deepened to the satisfaction of both of them.

## What a Relief!

For Jenn, a singer/songwriter and now a stay-at-home mom, aloneness, isolation, and insecurity that nobody would be there for her had been a lifelong fear. At first, the fertility struggle exacerbated the insecurity and feelings of aloneness, but in the end, it showed Jenn how far she had come.

When I asked Jenn if her treacherous path to parenthood had enriched her life, she answered without hesitation, "I'm not alone. I'm thirty-five years old right now, and it took me thirty-four years to recognize that I have so many people who love me and support me. They may not be loving me and supporting me in a way that I feel that they are understanding me at all times. But I realize that it's not always about understanding me. It's about being there for me."

She continued, "I had people praying for me and people going to Israel and taking my picture with them and putting it in the Wailing Wall. That was overwhelming to me, a beautiful thing that I got out of this. All those years that I felt 'I'm alone, and people don't totally get me'—that's gone. I will never feel that way again. People might not get me and there might even be times that Josh (her husband) is not going to get me, but it doesn't mean that he doesn't love me, and it doesn't mean that he's not there for me. It just means that he's totally not getting me *right now*."

When all was said and done, feeling that she should tell no one opened up wide into telling everyone without caring what people did with the information. Jenn became confident of the strength she found within—strong enough to know that if others handled her situation poorly, then *they* had a problem. She saw it as no skin off her nose because the payback in seeing the lengths to which other people went to support her and love her was stunning, awesome and, more than anything else, healing.

## Everything is "Relative"

Julie, a researcher, was very insular at first. She and her husband Dan were the only ones who knew what they were going through. She did not create a support network for herself. "There was nothing wrong with me, and I wasn't forty," was her attitude. Eventually, she began to suffer from the isolation and she chose to join my mind/body support group, relieved, she said, that she "wasn't the only one going through this." While she and her husband still had their secret from the larger world, at least she no longer felt that she was in a "vacuum." She conceived on the first attempt at in vitro fertilization.

Things changed drastically for Julie after she miscarried at sixteen weeks. It was at that point that she realized that she had to reach out to others. She told me, "I had to widen my circle and not be afraid to tell people that I was going through this. In my opinion, you cannot go through this alone. It's too physically and emotionally draining."

The interesting thing in Julie's case was that her opening up included opening up to her mother, a person who had not been on her trust list. Julie said, "It occurred to me that Mom should know, and I could tell her as long as I was clear to her that this is what I need and this is what I don't need." By taking this risk, a reconnection with her mother turned out to be a happy by-product.

Julie's mother learned to respect Julie's boundaries. Intrusiveness was the issue that had put her on the no-trust list in the first place. Her mom asked if she could tell one of her friends whom Julie knew and liked. Julie was impressed with her mother's intuition and agreed because she knew that this friend would pray for her. Mom's friend did pray for her, and also sent her books and letters of support. As is true with many in the fertility boat, a spiritual dimension had increasing significance for Julie, so by opening the wall of secrecy, she opened a lot more.

## Needing Privacy Can Swing Like a Pendulum

I have seen the privacy-secrecy issue swing in both directions. Some swear themselves to secrecy and if the months drag on into years, they throw all caution to the wind, tell the world, and feel relieved to have done so. Others tell everyone at first, perhaps because they have benefited from support in the past, or perhaps because they are optimistic about a speedy, positive outcome. But, if the infertility endurance test goes on beyond a certain point, some resent that everyone is looking at them with anticipation, often not knowing what to say. These couples pull back and become more insular. There is no right or wrong way to do this nor does your choice need to be set in stone.

> I'm the type of person who likes to hear from people who have gone through the same thing that I'm going through. Now that I've been through so much of this, I feel empowered to be the one who can say to people, "It's gonna be okay. You can do this." It makes me realize how far I've come. — *Stacie, teacher, New York, NY*

## Distribute the Burden

Opening up is largely about unburdening. Most people feel understood by others who are in the same boat. Yet, the medical setting brims with irony. Stacie, a teacher, captured the spirit of it when she said, "You spend so much time in the doctor's office, not knowing what's going to happen next, sitting next to people who are going through the same thing, but nobody's talking to each other, and it's isolating." Even among others who are out of the mainstream, an absence of connection with others in the same tributary can be painful.

People rarely make connections in the clinic waiting room. The medical setting is fraught with anxiety. If you can read body language you might know if anyone is open to conversation. Yet you may prefer to connect anonymously yourself. If so, Web sites, chat rooms, or blogs might suit you. Another way to reach into the virtual world from the comfort of your own home is to locate a teleconferencing event that meets your needs. Collecting information helps to reduce anxiety.

If you need live bodies to relate to, you can search the Internet and find local support groups, or reach out to friends or friends of friends who are veterans of the fertility pursuit. While support groups can be a real boon, they do not resonate for everyone. As Shari the event planner put it, "Before I joined your mind/body support group, I was in a general infertility support group. I found it distressing to listen to others bitch about their infertility. It made me feel worse."

However, others find infertility support groups—whether professionally or peer-led—to be invaluable. Participation in support groups does distribute the burden and lighten it for most participants. A support group can provide a user-friendly place in which to ease feelings of isolation, experience a safe haven, and pick up tips for effective coping, all in a cost-effective way.

Unlike the general support group that Shari spoke of, mind/body stress reduction groups are psychoeducational. Rather than focusing on airing pent-up feelings, they provide a setting in which you can learn skills that calm the mind and body. They are not therapy groups, but they are therapeutic. It is relieving to have specific ways of reducing stress at your disposal. *Most important is the fact that this approach emphasizes gathering information that is missing rather than feeling you are missing some screws in your head.* Learning mind/body techniques fills in the gap between how you cope and what you can learn in order to cope better. This makes it easier to get through this ordeal without feeling like a cork in the ocean. The majority of those interviewed for this book were former mind/body group members, or I taught them these skills in private sessions. Their enthusiastic participation in these pages is a testament to the liberation derived from choosing this method.

## Help!

Reaching out to others has many facets. There is the medical realm, the friends and family realm, and the live support group versus the virtual realm. And, there is the realm of individual- or couple-oriented professional help.

Are you one of those people who is reluctant to ask for help? Unburdening can feel particularly difficult if you've always been ferociously self-reliant. Maybe your brand of unburdening resulted from jogging a bunch of miles per week or working out in the gym. Has your doctor put a limit on vigorous exercise? Many women turn up in a foul mood because this outlet is blocked. Has meeting up with friends for happy hour come to an unhappy halt? Club soda doesn't cut it when your friends are imbibing in that glass of wine that is now off limits for you, not to mention that it's a dead giveaway that you're trying to conceive. Whether you pull away from your friends or change your behavior, they will suspect that something is amiss. Damn!

Joining a group is one thing. Reaching out for one-on-one help is a whole other kettle of fish. Some prefer it, but others, who struggle with feeling physically defective, rail against seeing themselves as emotionally defective as well. But listen up...*this is not about anyone's defects.* Infertility is a raw pain and difficult to fathom. It ranks way up there as a stressor to which people respond as if it were a terminal illness. You're in good company. Feeling crazy? You're normal.

The infertility struggle has a way of magnifying everything. This set of circumstances is highly charged. There is so much strength required of you. Given the arduousness of the treatment and the social, emotional, and spiritual fallout, seeking professional help, if you need it, makes sense.

Distributing the weight of the burden beyond trusted family and friends to a well-trained professional can matter a lot. If you choose to keep this information from your family and friends, at least you would have one outlet. A well-trained objective outsider can help you to monitor your *reactivity to* your circumstances. Infertility is a circumstantial stress. How you react to it, determined by your attitudes and beliefs, can make the situation worse. If you have someone who can help you see what you look like under duress, you can find a better way to *be.*

What are the barriers that might make it difficult for you to ask for help? Those in your world may have an attitudinal residue that presumes that seeing a therapist is a sign of weakness. To my way of thinking, it is just the opposite. It takes enormous strength to get beyond your own ego and admit that you, like the rest of us, have blind spots. Jogging, going to the gym, or meeting friends for a glass of wine cannot eliminate blind spots; it can only distract from them. If jogging as an outlet is behind you, connecting with your strength and inner resources that you might not realize you have is in front of you.

> I keep thinking if I had come to you two years earlier, maybe my struggle could have been shortened. — *Ellen, photo editor, mother of twins, New York, NY*

## Has Your Core Inner Strength Run for Cover?

It's not that you're not strong. It's that you need to feel access to that strength at the same time that you want to crawl in a hole.

Infertility is expert at creating a sense of weakness. Stacie, the teacher, told me that she had gotten to the point where she felt that her inner strength had "eroded away." By reaching out for help, Stacie reclaimed her capacity to assert herself. The assistant principal of her school objected to her taking a few days off when she was about to have an embryo transfer. Stacie told her to report her if that's what she needed to do, but she'd be taking the days off anyway. The assistant principal backed down when Stacie brought her strength forward.

When so much is at stake, finding your strength is mandatory. Getting help can be particularly useful in standing up to those people, places or things that need to change. Donna was able to confront her husband's heavy drinking. Shari and Ruth were able to command more sensitivity from their husbands. Ruth and Nanci were able to set limits to intrusive grandparents. Brenda was able to assess that geographic distance from the "firing squad" was "healthy."

It is wonderful when awareness grows about just how strong you really are. As Jenn, the singer/songwriter, told me in a private session, "Here I am, my diagnoses had diagnoses: DQ alpha,[18] a clotting disorder, antiphospholipids,[19] four miscarriages, a uterine lining that got messed up from a D & C, which then had to be redone followed by an intrauterine infection. I'm taking seven shots a day, worried about what was going into my body, and I realized that I was really strong to go through all of that for two years."

Awareness of how devastating it is to be in this situation may override access to your strength. Anxiety-relieving guidance from professionals is a phone call away.

Besides a bias against seeking help as a sign of weakness, children in some families learn that to go public with any distressing situation is equivalent to airing the family's dirty laundry and is considered taboo. Of course, the twist here is that baby making should be a private issue. But the right kind of professional help has the potential to bring you from feeling victimized by these unasked for circumstance to feeling vanquished as you meet this challenge and grow from it. *To feel back in control—not of the situation but of how you handle it—can result in a very pleasant surprise as you get to know yourself in new ways.*

## Creating an Eye and an "I" in the Storm[20]

There are many metaphors for the infertility experience such as roller coaster, merry-go-round, battlefield, nightmare, and storm. Take your pick. Notice that they are all about dizzying movement. This is not conducive to the quietude and stability that fosters the self-reflection we need when faced with challenges. For any of us, to be able to think straight when under duress is hard enough, but for this to happen while in a spin adds another layer of difficulty.

It is useful to look at the storm metaphor here. A hurricane is a large, swirling vortex where torrential rain and gale force winds can devastate. Yet the center, the eye of the storm, is a place of stillness through which the sun shines. You can get an idea from this picture of a hurricane from space. The eye is about two hundred miles wide.

In nature, the eye of the storm happens. With infertility, an eye in the storm must be created. . .and recreated. The inner stillness we need, particularly to engage in letting-go coping, is unlikely to happen by itself. The need for a place of respite is great and not easy to achieve. Yet, you will be better qualified to deal with the built-in challenges of infertility when you find an inner oasis from which to sort through choices and make good decisions. The exercises at the end of chapters 7 and 8 will guide you to what serves this purpose.

The exercises at the end of this chapter will teach you new ways of seeing yourself and new ways of being. The question in this chapter is Who needs to be there with you?

Connecting with others is an important and relieving part of creating respite. But, a place of respite is also a place from which to create connections with oneself, since it is so common for one's self-esteem, one's "I," to get shattered on this journey. This chapter is about reaching out. This book is about reaching in.

## Exercises for Building Assertiveness

Assertiveness is an important skill. Are you someone who behaves as if *no* is a four-letter word? If so, it could hamper your ability to set the limits with people who, wittingly or unwittingly, intrude or expect things from you that you need not feel obliged to provide. Here are two exercises that can curtail saying yes when you mean no.

1. This is a problem-focused exercise to build confidence for assertiveness:
   Practice saying no in a safe environment as a game with any friend or relative who you trust. Have fun with it. For instance, if your partner says, "Will you make me hot dogs for dinner," say, "No," even if you intend to. If your friend wants to take a walk with you say, "No" with a twinkle in your eye. Make it up as you go along.

2. This next exercise is an imaginary rehearsal. Use the power of your creativity to visualize a person to whom you must say no for whatever reason:

   1. Find a quiet place where you will be undisturbed.
   2. Take a deep breath and close your eyes.
   3. Remember that the limbic system of the brain does not know that you are not where you imagine yourself to be. This is important. You can believe that by rehearsing saying no, it's as if you already have.
   4. Anticipate what the other person's response will be and what you will say in return. If you are having trouble with this, imagine that you are the size of a giant or the other person is a miniature version of him or herself.
   5. Remember that the bottom line is no. You can end the rehearsal with something like, "I'm sorry that you feel that way, but I will not...(fill in the blank).

## A Breathing Exercise to Build Self-Connection

Breathing exercises are easy to find in books. They are also easy to make up, such as breathe in to the count of four, and breathe out to the count of eight. This exercise is different. It takes into account how, in this society, we tend to breathe shallowly, as if we fear an imminent danger.

Here's how to reverse that inclination:

1. Sit up straight. Form a circle with your lips.
2. Slowly, very slowly, begin to inhale. Continue to inhale for as long as you can, keeping your neck and shoulders relaxed and barely moving.
3. Release the breath just as slowly.
4. Do this with a clock with a second hand with the intention of extending the time it takes to complete a full breath.

By filling your lungs to their full capacity, you are retraining them to remember the joys of their full expansion. It is also an important way to symbolically create connection with yourself by tuning into your life force.

Chapter 5

# Depression and Anxiety: Finding Force in the Life Force

Problem solving coping and letting-go coping are both important skills to have in the battle with infertility. Problem solving coping is primarily about making behavioral changes, what I've called *doing*. Letting-go coping is about using one of several techniques by which you can separate yourself from the turbulence of infertility (behavioral disengagement). In the process, you can break the spasm of stress that grips mind and body so you can just *be*.

While this is a simplified way of looking at coping options, things are not so cut and dry. Sometimes doing something allows you to be more at peace, like doing yoga, for instance. In other words, doing can serve to clear a space in which you can just be. Conversely, letting-go into being can clear a space that will allow you to do something helpful for yourself.

This chapter elucidates the doing and being coping styles of Sarra and Kim, women who struggled in their own way to master the predominant and powerful emotions that manifest for infertility patients, namely depression and anxiety. You will see what Sarra "did" to win the battle and the way that Kim learned to let herself "be." These stories are meant to energize and fortify you for the endurance test and help you to find the force in the life force within you.

## Utilize the Energy of Upset

Depression and anxiety come with the territory. How could they not? Apart from the small percentage of the population worldwide over the history of man on earth who chose not to have children, for the rest of us, having children is a biological imperative. Unlike other members of the animal kingdom where the instincts and the senses drive procreation, ours is a total mind/body imperative and as such, the heart is involved. For humans, there is a yearning that goes along with the capacity for sperm and eggs to connect and a worry that they will not. Depression and anxiety feel dreadful. Unfortunately, both of these states are waiting to take up residence within you.

Depression can range from sadness to physical and emotional immobility. The span of feelings for anxiety varies from feeling unnerved to wanting to jump out of your skin. The whole gestalt of infertility can be overwhelming especially because of the sick feeling one would get in the pit of one's stomach when there is no sense of when or how this challenge will end. How anxiety provoking and depressing is the fear that this nightmare will never end or will end badly? At the outset, it's hard to imagine a way back to normalcy, never mind how to utilize your pain to change and grow.

## Popeye and Depression

I believe that when it comes to infertility, there is an aspect of the life force that kicks in, which I would call "the Popeye effect," borrowed from the cartoon character. When Popeye's life force needed shoring up, he guzzled down a can of spinach and his biceps bulged. The message to the children watching the cartoon is, spinach may be distasteful, but it will make you strong.

Can you think of the energy contained in the drive to reproduce as joining forces with the energy of depression and anxiety, shoring up your inner strength? Can depression and anxiety fuel the life force as if it were Popeye's spinach? Can this be the fertility in infertility?

This is not to say that the mind and body do not experience the weight of these emotions as delineated in psychology books. Joy in life might be elusive, sleep patterns may be disturbed, eating patterns may become a salve or an aversion, socializing can be burdensome, and the prospect of navigating this maze can feel next to impossible if you cannot seem to settle yourself down. The question is can you harness these emotions to outpace these symptoms?

## Sarra Says No to No

Sarra, an advertising executive, shares her story, inviting you to leave leeway for optimism and hope if you feel trapped in depression. After her second IVF, the doctors were sounding discouraging and speaking to her about her chances of conception in very low percentages. She was thirty-nine years old. She was told that she had poor egg quality and was a low responder to the stimulating medication. Her FSH became elevated.[21] Of the few eggs that she produced, fewer fertilized, and none made it to transfer. These were the medical facts. She became extremely depressed.

Sarra was able to rise above the depression to a place that activated the "fight" in her. She felt intuitively that her mood was as much a consequence of her dismissal by the medical community as it was to the medical facts themselves.

She became preoccupied with worrying about what her life would be like if she could not have a child. As her mood worsened, she told me that she came to feel that there would be "no point in living" if she couldn't have a child. At first, she was resistant to taking the antidepressant medication that was prescribed for her. At the same time, she knew that she was isolated and that she needed to reach out for help. When I interviewed her for this book she told me that she had realized that the "infertility mountain was too high to climb" by herself, and she joined my mind/body support group. Soon she began taking a mood-elevating medication, too.

With the lift from the group and her medication, Sarra was able to step back connect with the energy of her upset. She felt fueled to follow what her intuition was telling her. She said to herself, "I want to have a child. What if the facts are not being interpreted correctly?" Simultaneous with feeling seriously depressed, the Popeye effect was operating. As you will see, her infertility created a kind of fertility akin to grass that grows through tarmac. Sarra told me, "I was not going to give up without a fight." In the midst of a devastating depression, she said, "I got fed up with this whole medical community telling me that I had no chance…It was out of my set of comprehension that they were playing God with me." She became enraged that they showed her "no humanity or compassion." Her capacity for anger helped her to rise above the depression and became the spinach or the fuel. What the can of spinach is to Popeye, the force of the life force to reproduce can be to the infertility struggle. Her determination to persevere began to run parallel with the debilitation that had rendered her unable to function adequately without antidepressant medication.

When I interviewed Sarra for this book, her daughter was six weeks old. She told me how raw her emotions had been. It is true that the medical community does have statistics, and technically, she fell in their category for abysmal odds. But she preferred to focus on those with lousy odds who succeed. Sarra mobilized herself. She agreed to take psychotropic medication. She joined my mind/body group, which addressed the feelings of isolation and gave her much needed support and coping tools. She was available and receptive to learning what to do and how to be.

Especially under these circumstances, it is unreasonable to expect depression to disappear with medication. But for Sarra, her depression was greatly eased. She was free to apply brainpower and evaluate those who were evaluating her. She felt that perhaps there was medical information that the doctors were not considering. She kept saying, "This does not make sense," words that showed a capacity to think about what her doctors were telling her. She was prepared to challenge her doctors, displaying a capacity to be assertive. *This is not what you would expect with severe depression, so if you are in Sarra's category, can you find your fuel?* She was on a scientific quest to find doctors who thought "outside of the box." She told me that she wanted a medical team "who would be willing to work with [her] and not against [her] by just saying, 'You're too old.'" She was also on a quest to feel as understood by the medical community as she felt with me and with the group.

To make a long story shorter, Sarra was relieved to find a clinic—quite a distance away, mind you—that respected her research and did think outside of the box. Sarra had unearthed an approach that had to do with immunology. Her new doctor supported her wanting to pursue immune treatments even further away—in Greece, her native country—where a cousin with an immune problem had been treated successfully. Sarra had two treatments with Leukocyte Immunology Transfer[22] and conceived—without in vitro fertilization, no less! That turned out to be a biochemical pregnancy, but she was heartened. It felt to her as if her system had been jump-started.[23]

The immunology approach was quite common in 1979 and in the early 1980s, when I first began my private practice, and then it fell out of favor. Although it is still not standard practice in the IVF clinics at this writing, and is controversial, the immune approach is back in the story in some clinics. Please keep in mind that I am simply telling Sarra's story. I am not in a position to evaluate it.

What is most important was that immune treatment made sense to Sarra. Her research then led her to a facility in California that worked with her clinic in the New York area. The blood work ordered by the California clinic was different from any she had had, and the results confirmed her intuition that hers was an immune problem. She had a new sense that there could be a light at the end of the tunnel.

Meanwhile, Sarra told me, "At this point I began to have something to look forward to. I was feeling more confident, and my psyche was in much better shape than it had been for the last two or three years. I started to do things that I had enjoyed in the past." Most importantly, she said, "I started finding myself in bits and pieces. I wasn't as whole as I used to be before this infertility had started, but I was finding a little Sarra here and a little Sarra there."

Meanwhile, her immune story was unfolding. It was determined that her natural killer cells were very elevated. She was presented with a solution. The treatment was the same as that for an autoimmune disorder such as rheumatoid arthritis. Her spirits were totally lifted.

For purposes of conception, she switched antidepressant medication from Prozac to Lexapro, because she was told that Lexapro "provides better neurotransmission to the uterus in addition to providing antidepression properties." Now she felt as if she was with scientists who knew about *her* specific body chemistry as well as biochemistry. Sarra conceived without assisted reproductive technology and carried her daughter to term. She gave birth at forty-two after a three-year quest.

This was a wild goose chase that involved educating herself as if she were pursuing a medical degree, making complicated choices with her husband and doctor, and traveling far and wide for her happy ending. This defies the definition of depression, does it not?

In the aftermath of her nightmare, Sarra wondered why her original medical team could not get past the facts of her age and her low response to *their* protocol and see her as a person. Even though her doctors were right statistically, Sarra wished that they could have looked at all possible protocols.

It was unnerving to Sarra that some reproductive endocrinologists do not realize the level of emotional agony associated with infertility. In her words, "We are really raw and the emotions come from deep down—from the bottom of your feet." She wanted to feel compassion from her doctors given that "the whole idea of infertility for [her] was a cloud on top of [her] head...something that [she was] dealing with on a daily basis." No one on her original team led her to believe that there might be a solution beyond ovum donation or adoption. Sarra, despite a profound depression, made her way to the place where hope could reside—the community where others were conceiving and where the odds were stacked against them, too. It may seem counterintuitive to think that strength and vulnerability can coexist, but they certainly can.

I realize that it might be hard for some of you to read Sarra's story. What if you do not have the funds to pursue every avenue? Worse, what if you do have the funds and you still do not achieve a pregnancy? This gives me another opportunity to say that for one thing, you would still have options to achieving parenthood, although I understand that you and your partner would need to shift gears.

But, what if you couldn't afford those options either? I've seen people come up with some wildly creative solutions for this, driven by the Popeye effect. Parents often step up to the plate. In that case, it might take some effort to rework your relationship with your parents in order to ask or accept such a loan or gift, but that is part of the opportunity for growth. And, there are scholarships you can apply for at www.INCIID.org and perhaps other organizations. Some clinics have payment plans.

What if you can't or don't want to see your way clear to any of the above options? Would that mean that you need to accept that you'll be living child free? At this stage in your quest, you may not need to answer the question, "When is enough enough?" For now, it might be best to access the Popeye within you, keep your eye on the prize, and not expend energy on a possible future that may not unfold.

## Popeye and Anxiety

Anxiety has various characteristics in common with depression. Vulnerability provokes both. Both are mind/body experiences and can be seriously uncomfortable. You can reframe both as gifts because they grab your attention, giving you a chance to figure out what needs to change.

Like depression, the energy of anxiety can be fuel or "spinach" for the fight. In fact, since anxiety is a revved-up feeling, going into action (doing) can not only *feel* palliative but can also *be* palliative if it discharges tension and brings the body back into balance, even if only temporarily. That is, back into balance *unless* denial of what is going on goes too far as it did with Kim.

## Kim says Yes to Yes

For Kim, an advertising maven, the intensity of anxiety resulting from a failure to conceive after two years of trying was what finally made her realize that something had to change. Until Kim's self-awareness grew, she was as a gerbil determined to keep her wheel spinning nonstop.

Kim did not experience anxiety as anxiety per se. She was only aware that she "felt stressed." She reminded me that she "had built a high-powered career in advertising" and "lived stress…without knowing any better…ignoring the obvious signs." Looking back, she admitted, "How I ever thought that working twelve-hour days, being pulled in ten different directions by clients, bosses, and colleagues, and then schlepping across New York City sitting on a loud, dirty subway train and falling into bed at midnight was okay is beyond me. I never stopped to listen to my body…and how powerful an enemy stress was in my life."

When Kim joined my mind/body group, she said, "My eyes opened to what my life had become." She went on to explain an evolution in which she learned to embrace various antidotes to stress. Stress for her was *un*experienced anxiety. Kim had used a Popeye-like stubbornness to keep the inconvenient feeling of anxiety at bay. When she gained sufficient self-awareness, she shifted that stubbornness so that it could work for, not against her. She used the Popeye effect, her determination and mental muscle, to make up for lost time by changing her ways.

As a member of my group, Kim learned the relaxation response, which you learned at the end of chapter I (and I hope that you are practicing!). She told me that "doing the relaxation response at first felt foreign. It was very difficult in the beginning to bring about a change. I had been in a mental frenzy for twenty years . My mind and body literally rebelled against [achieving serenity]. I would put it off and procrastinate. But every time I did the [relaxation] practice in group, it felt amazing, so I would force myself to do it at home. At first I couldn't make it more than three minutes without falling asleep (yeah, my body was telling me something). So I began to do short meditations in crazy places—on the subway, at my desk. Even walking across the city, I could do the mindfulness walk that I had learned. I realized that there were plenty of opportunities to focus on taking better care of myself that I had never taken advantage of."

Kim went on to explain, "After two surgeries, one miscarriage, and seven fertility treatments, I decided to make big changes. With the support of my husband and the [mind/body] group, I found the courage to tell my boss that I was going to take six weeks off to do another IVF cycle. He was not happy, but I didn't care." (Notice the value of assertiveness, ladies and gentlemen.)

In addition, she told me, "I cleansed myself of the craziness of advertising and walked into the next cycle with a whole new attitude. I spent mornings in Central Park and meditated under the trees. I began to notice beauty all around me. I found a new appreciation for life. I began sleeping better and eating better, not because I was trying, but because I felt better about my life.

And I got pregnant and had a beautiful baby girl! Never again will I allow life to run me. I have achieved balance. I am at peace." Two years later, Kim produced a second daughter, conceived as "a freebie" and delivered at age forty-three without a hitch.

## The Takeaway...

Sarra and Kim's stories illustrate the potential to convert paralysis into action when you feel stuck in emotions as powerful and profound as depression or anxiety. When it feels as if you are stuck in depression or anxiety, it is important for you to see that where you are really stuck, is in yourself. Seeing yourself clearly so you can learn to be different is what most needs to be changed. When you can *be* different, you can *do* things differently, and vice versa. In the process, you can keep hope alive. But at the same time, keep in mind that it would be unreasonable to expect yourself to navigate through this challenge without experiencing a good catharsis from time to time. You're only human.

# Exercise

## Cognitive Restructuring

This exercise will help you to rework the way that you think. When negative thoughts automatically intrude, it's tempting to believe them as if they were gospel truth. Dr. Daniel Amen is credited with saying, "Don't let the ANTs (automatic negative thoughts) spoil your picnic."[24]

You can subject these ANTs to scrutiny. It may surprise you to learn that the feelings that go with negative thoughts are *not* facts. That's right, folks, you may feel trapped in negative thinking, but you have the power to replace what zooms into your mind with a better way of framing experience. You can *decide* if you want to believe these intrusions. And you can go one step further and learn how to convert them into affirmations. You are not condemned to feeling overpowered by depressing or anxiety-provoking thoughts. This exercise can be to you what spinach is to Popeye. Learn the technique. Then you must discipline yourself to flex this new "muscle." If you do not work this exercise, it will not help.

Here's the technique:

1. Turn your attention inward by closing your eyes and taking a deep breath. Pay attention to the rhythm of your breathing. Use the breath to settle yourself down. If you're feeling particularly agitated, do the relaxation response long enough to feel a shift.

2. Let your mind wander until you remember one or more negative thought(s) that come to mind without your permission. Open your eyes and write them down. If you've come up with more than one, save the other(s) to work with at another time. An example of an intrusive thought might be, "This IVF will never work."

Write the ANT(s) here:______________________________________

______________________________________

______________________________________

3. Close your eyes again and ask yourself where in your body does the intrusive thought makes its presence felt. For example, perhaps your negative thought makes you feel heartsick, or you feel it in your gut. Maybe you feel pressure in your head or your throat feels constricted.

Make note of the *body* experience here:________________________

4. Next, identify how you experience the negative thought in your *mind*. What mood predominates? Do you feel sad, angry, confused, or a combination of feelings? Check out the list below of the variation of eight adjectives. Write down all of the feelings that apply.

| | |
|---|---|
| Angry | upset, bitter, resentful, irritated, annoyed, indignant, hostile |
| Depressed | discouraged, guilty, pessimistic, despairing, sulky, powerless |
| Confused | bewildered, frustrated, disillusioned, skeptical, lost, unsure, pessimistic |
| Helpless | incapable, tragic, paralyzed, alone, vulnerable, dominated, alienated |
| Indifferent | numb, bored, lifeless, insensitive, |
| Afraid | terrified, nervous, worried, wary, frightened |
| Hurt | crushed, tormented, dejected, aching, victimized, offended |
| Sad | tearful, grieved, pain, anguish, lonely, desperate |

Make note of the *mental* experience(s) here: ____________________

________________________________________

By now you should have mind and body labels for the automatic negative thought.

5. Once you have realized how your body and mind experiences the negative thought, you have a chance to scrutinize and invalidate it. You are prepared to look at your intrusive thought through the filtration system of the questions that follow. *The idea is to start with the presumption that the thought is a distortion, or irrational in some way, even if there is a grain of truth in it.*

Write the automatic negative thought here again __________________

________________________________________

and ask:

a. Is it really true that (this IVF will never work, for example)?
______________________________
b. Am I using the past to predict the future?
______________________________
c. What is the evidence that (this IVF will never work, for example)?
______________________________
d. As a rule, do I tend to exaggerate the negative? __________
e. What would I say to a friend who was in this situation? ________
f. What would happen if I eliminated the words *never, ought* or *should* from the sentence? ______________________
g. How does it help to honor the negative thought by believing it?
______________________________
h. What if I choose *not* to believe the negative thought? ________
i. Am I rejecting the possibility of an outcome that is different from the negative thought? ______________________
j. Am I having a pity party?______________________
k. Am I blaming myself?______________________
l. If I wanted to be kinder and less judgmental toward myself, how would I change the sentence? ______________________

Using the answers to the questions above, begin to ponder how you would *reframe* the negative thought. For instance, would you say that *because* IVF has not worked until now, you have had a chance to save more money or solidify your relationship with your partner? Perhaps you got closer to your mother or a friend, or you came to understand what a lovely, sensitive person your boss is?

Take your time with this. It's not easy to think out of the box when you've been locked inside of it. It might help if you reworked your negative thought with a trusted friend or a therapist.

After you've processed these questions, and in preparation for reframing the negative thought, ask yourself if you are willing to let go of rigid adherence to the thought.

6. Now close your eyes again, take a few deep breaths and find a reframe. You may need to launch yourself with a few minutes of the relaxation response.

Write your reframe here______________________________

7. Then, trust that you conscious and unconscious minds will team up to find a way to flip the negative thought into an affirmation. "This IVF will never work" can morph into "I'm doing everything in my power for this IVF to work" or "My medical protocol is different, so this time the IVF has a better chance of working.

8. Write the flip side of the ANT, now an *affirmation*, here: ________
________________________________________

9. Now ask yourself Where in your *body* you feel the affirmation?

Write the *body* experience here:
________________________________________

10. Which of the following pleasant "feeling" words go along with the *mental* experience of the affirmation?

| | |
|---|---|
| Positive | optimistic, hopeful, relieved, confident, inspired, excited |
| Strong | hardy, secure, confident, free |
| Loving | sensitive, devoted, close, passionate, touched |
| Happy | joyous, satisfied, cheerful, great, lucky, delighted |
| Alive | animated, easy, liberated, optimistic, spirited wonderful |
| Open | understanding, easy, interested, receptive, kind, accepting |
| Interested | fascinated, intrigued, absorbed, curious, concerned |

Make note of the *mental* experience here:
________________________________________

Compare the difference in the mind/body experience of the ANT with the mind/body experience of the affirmation.

Once you've created your affirmation, you can use it as a phrase with the relaxation response. And when you learn self-hypnosis at the end of chapter 7, you can use it as a posthypnotic suggestion.

It is worth repeating that you must discipline yourself to flex this new "muscle."

Chapter 6

# Up Close and Personal: Entering the Bodymind for Effective Coping

Stress cannot discriminate in its impact on mind and body because mind and body are not separate. Think of it this way: the physical, psychological, behavioral, cognitive, relational and spiritual symptoms provoked by stress that you discovered by taking the self-rating scale at the end of chapter 2 are the body's wisdom. The mind sends its confusion, upset, worry, or fear into the body in the form of tense muscles, headaches, bellyaches, or elevated autonomic functions, such as heart rate and blood pressure. Conversely, whatever the body experiences makes its way into the mental sphere, resulting in any of the myriad symptoms listed in the stress warning signals chart on pages 23 and 24. Because of the mind/body unity, the impact of stress travels in both directions along a two-way street. This is how we know what is going on in our world.

Skeptical? There is a simple way you can experience being alive in your body and mind. Try stamping your feet hard a few times. Then close your eyes and experience the way you feel the sting and the ongoing vibration that travels up your legs. The only way to experience this is in the present moment. Experiencing the mind/body unity in the "now" is as simple as that. Depak Chopra said it even more eloquently in a lecture I attended recently: "If you need evidence that the mind and body are connected," he said, "wiggle your toes."

## Understanding Body Language

If you can accept symptoms as mind/body wisdom that flows from mind/body unity, then you'll notice that the fertility quest presents you with one wake-up call after another to mind and body awareness. If you're like Kim in chapter 5, physical and emotional symptoms can hit you over the head, to no avail. Kim dismissed the stress as external to herself. It took two years of struggling with infertility before she began to respect stress's capacity to alert her to the need for change.

If we mentally dismiss awareness of how we feel, the body will hold the information anyway. No amount of denial will banish our "body of knowledge." Symptoms are our body language desperately trying to "talk" to us.

When Shari, the events planner, was a member of my mind/body group, she had become aware of the connection between the stress of the journey and exacerbation of what she called her "major stomach problems and acid reflux."

Unlike Kim, Shari allowed her symptoms to register as a plea for help. When Shari was hoping for a conception, she allowed her digestive woes teach her that something needed to change. After three intrauterine inseminations, three hystosalpingograms,[25] surgery for endometriosis,[26] a miscarriage from an unexpected spontaneous pregnancy, two fresh and one frozen IVF cycles, and a botched D & C, it was a relief to say, "Enough!" Shari realized that she wanted a family more than she wanted a pregnancy, and she and her husband, Steve, resolved their infertility with the adoption of their daughter, Mai. All through that harrowing process, Shari integrated mind/body techniques into her repertoire, healing the experience of infertility and at the same time learning to mitigate her pattern of putting upset into her gut.

The condition of infertility is one thing. The medical treatment of it ups the ante in the stress department. When you're on the treatment battleground, you're bombarded with these crazy, complicated words that go flying completely over your head like whizzing bullets. Words like intracytoplasmic sperm injection,[27] sperm DNA integrity assay,[28] amniocentesis,[29] and comparative genomic hybridization[30] all make you feel like waving the white flag. Then you find out that you'll be taking Lupron, a drug that mimics menopause. You're here to get pregnant and they're inducing menopause? What the hell? Eventually, you understand the logic and value of all of this. In the meantime, *your body responds as if you are actually on a battleground.*

> I became aware that my body was telling me that I was not handling stress well. I was not ovulating. My hectic lifestyle and job were not compatible with the centeredness and focus that healing from my amenorrhea would require.[31] I decided I would make extreme self-care my day, evening, and night job. — *Melissa, artist, Dallas, Texas*

## The Face-Off with Stress

Adversity happens. This is life. We need to develop methods of self-soothing and managing the stress of adversity, otherwise known as coping.

No matter how haphazardly, we learn to function over, under, and around, if not through, stressful life events.

Yet inevitably, life brings us to areas where coping skills fall short through no fault of our own. Challenges might have been too big when you were too small, setting you up to feel inadequate when faced with life's trials and tribulations. Perhaps you were traumatized early on and you freeze now in the face of any emotional challenge. Perhaps your caretakers were emotionally handicapped themselves and, therefore, would have been insufficient role models and educators. It's hard to learn to cope effectively if no one shows you how. If you were shielded from stress altogether, you might be at even more of a loss when it comes to coping.

> I can't even describe how much of a struggle [it was] to feel so sad and lonely and depressed and horrible…until I learned some of the mind/body coping techniques. They helped me move from feeling myself to be a failure…to focusing on what I could succeed at. — *Shari, former events planner, current yoga instructor, New York, NY*

## "Plop, Plop, Fizz, Fizz...Oh What a Relief It Is"

Mind/Body techniques are like Alka Seltzer for the soul. Monique learned at first through yoga and then through other mind/body methods, that "one second of peace" could get her "addicted" to feeling relief. For her, the empowerment that followed suit was a remedy for feeling swamped by stress.

Near the outset of their infertility struggle, Monique, an MBA/lawyer, and her husband Bill, a financial wizard, were dealing with what they called the "trifecta" of stress. Monique had had a molar pregnancy, a cancerous growth that needed treatment, separate and apart from losing that baby. In addition, time had run out on her congenital heart defect. She was facing open-heart surgery. Also, she and Bill had been working for Enron when that scandal caught them in the fallout. Stress for them came in triplicate and kept magnifying exponentially.

When all was said and done, Monique had seven miscarriages—four with IVF, two from spontaneous pregnancies, and the molar pregnancy. She credits the birth of their son to a New York City practitioner who was like Sherlock Holmes. He noticed what had been in her telephone book-sized chart for years but was overlooked by her previous doctors. She and Bill were passing urea plasma, a bacterial infection, back and forth through sexual intercourse. This infection is reputed to interfere with holding a pregnancy. Monique and Bill had been treated for this disorder with antibiotics but not simultaneously, as is the protocol. Of her five-year journey Monique says, "I got through college faster than I got my son."

Monique joined my mind/body support group. She said she joined the group "to address the need for human support." But, in the process, she told me, "The many mind/body options, taught me a lot about myself, what makes me tick and what *changes* I *have* to make to be happy. *The more I realized that infertility did change my life, the more it made me take steps to change my life.*" Monique actively sought changes that would give her mastery over her mind and body responses to stress.

During the long process, Monique was jolted into self-awareness, not only because of the infertility, but also because of her congenitally defective heart valve. She said, "When I found out that I needed open heart surgery, I needed to do something, so that I was not completely miserable all the time. This is why I started diving so deeply into the mind/body aspects."

She went on, telling me, "I had always done yoga, but I needed to make it a daily part of my life. When I did, I got that one second of peace, and that's what got me addicted. I got serious about yoga, nutrition, and all different kinds of mind/body work. Restorative yoga, meditation, guided visualization—everything. I was trying to do something at least once a day. I realized that some days, watching my breath did not work. Some days, I needed a walking meditation. Some days, if I was too scattered, listening to a tape really helped. I picked the tool for the day to get relief from feeling crazy. And I would say that therapy became one of my tools too." Living without joy for a long time created an awareness in Monique that all can change when we decide to pursue happiness.

> I would say the overall gain from learning mind/body techniques is that I feel more comfortable in my own skin. I trust my gut. Doctors were telling me to adopt, but I knew I could get pregnant and carry to term. I learned to trust that. — *Lauren, former lawyer, now stay-at-home mom, New York, NY*

## The Panoramic View of Mind/Body Coping Techniques

Coping means dealing with and overcoming problems. Two related and interesting definitions of coping suggest the kind of adaptation that we need in order to cope effectively. In one definition, coping describes the round-edged tile that gives a bathroom or kitchen backsplash or a swimming pool a finished look. In the other, coping refers to the strips of fabric that go between the sections of a quilt that help to make up for imperfections. Infertility leaves us feeling unfinished and suggests that there is something more that we need to recreate a sense of wholeness. We need to be able to ease the tattered edge of the mental and physical disruption caused by this unwanted reality.

**In lieu of an exercise at the end of this chapter**, below you will find a compilation of the full panorama of mind/body coping skills. They are considered mind/body because of the two-way street phenomenon—what eases the body eases the mind and vice versa. This list is organized to take into consideration the difference between doing (*problem solving*) and being (*letting go*).

I hope that by now, your level of self-awareness has expanded by:

- learning the relaxation response at the end of chapter I
- exposing lifestyle and other issues which may be adding to your stress through the four self-rating scales at the end of chapter 2
- practicing the communication exercise at the end of chapter 3
- trying the "imaginary" rehearsal for asserting yourself by saying no at the end of chapter 4
- using the cognitive restructuring to convert automatic negative thoughts (ANTs) into affirmations at the end of chapter 5.

By studying the various coping mechanisms below, when you need to deal with yet the next challenge that infertility tosses at you, you'll only need to stop and take a tranquilizing breath, and reach for this list.

## Coping by Doing

- Change a lifestyle habit.
- Change a family or work routine.
- Change a responsibility.
- Change a schedule.
- Change a communication style.
- Notice stressors at home, at work, or socially.
- Make a pro-con list.
- Clarify values.
- Evaluate relationships.

- Set goals.
- Evaluate your needs.
- Ask yourself what would rebuild your self-esteem.
- Distract yourself with something that is predictably pleasant.
- Monitor and shift thoughts to avoid being trapped in negativity.
- Reframe the situation in a better light. (How is the glass half full?)
- Allow a catharsis. Our tears are a safety valve.
- Reach out for social support.
- Find a way to laugh! (This has been studied; laughter *is* medicine!)
- Release emotions into a journal.
- Recognize and release judgment of others and, especially, judgment of yourself.

## Coping by Doing *and* Being

- Engage in exercise, yoga, and/or tai chi:
    - to feel physical strength and relax the body, and
    - to release endorphins and relax the mind.
- Release tension with a massage.
- Participate in energy healing, such as acupuncture or reiki.
- Learn relaxation techniques.

## Coping by Being

- Accept the reality.
- Focus mindfully on the present moment.
- Breathe purposefully and generously.
- Create affirmations and repeat the positive self-talk in your mind.
- Experience your spirituality.
- Flow with guided imagery or hypnosis tapes.
- Learn self-hypnosis and other letting-go techniques.
- Enjoy music, dance, singing, chanting, and/or drumming, all of which reverse stress physiology.
- Become absorbed in art projects.
- Play.
- Read or listen to stories.

You may be annoyed that you are being guided to build awareness, while thinking you'd rather live your life without having to study it. But like it or not, ever-increasing awareness is the name of the game.

## You Can Win at the Change Game

I'm well aware that choosing from the above list and following through on making changes is likely to be met with internal obstacles. The rest of this chapter will give you a leg up on getting beyond that part of each of us that is resistant to change.

There are two explanations for this resistance. I've alluded to both of them in earlier chapters, but here is the perfect place to take you deeper into the answers. Armed with this information, you will find it easier to commit yourself to making the changes that will empower you to cope with infertility more effectively.

## A Little Science on Our Resistance to Change

In chapter 1, I introduced a bit of brain science so that I could talk about the influence of our limbic system. I'd like to develop the concept here so that you understand your ability to rise above the demands of imprinted habits and fears. This way, the weight of your history will be less likely to pull you back to a place where you don't want to be.

To recap, the mind/body unity is a two way street. So whether you go through the "body door" to reduce stress with yoga (for example), which calms the mind with focused attention on the body, or whether you go through the "mind door" to calm the body with mental focus on the breath (for example), you come to the same vestibule: the limbic system of the brain.

The limbic system contains structures such as the amygdala, the hippocampus, and the hypothalamus. These structures hold imprints of early life experiences of trauma and account for our emotional reaction to stress. They are also imprinted with our current life stressors—hello, infertility. We must wrestle with these imprints if we want to make relieving changes.

And wrestle we can, thanks to what the emerging sciences of neuroplasticity and neurogenesis is teaching us! Without getting too technical, these terms describe the flexibility of our brains—just the opposite of the adage You *can't* teach an old dog new tricks.

Remember, the limbic system "learns" what's going on in our lives from the brain stem and the neocortex. Remember, too, that the limbic system cannot tell time. So if you close your eyes and breathe easily and regularly, as if all is right with the world, the brain stem signals the limbic system that it is safe to relax. If simultaneously, you're imagining that you're somewhere out in nature, the limbic brain does not know that you are not where you imagine yourself to be and it has no idea whether you're enjoying nature for ten minutes or ten hours. It responds as if you are there, because of your investment in the present moment.

To engage the brain stem and the neocortex with this kind of guided imagery is to speak the "limbic language." The relaxation response, hypnosis and other methods of behavioral disengagement also allow us to intervene in the stress imprints of the limbic system.

A two-paneled *Far Side* cartoon says it all. The first panel is titled What we say to dogs. In it, the master is scolding his dog, Ginger. He's saying, "I've had it, Ginger. You stay out of the garbage! Understand, Ginger? Stay out of the garbage, or else!" The second panel, titled "What they hear" reads, "Blah, blah, blah, Ginger, blah, blah, blah, Ginger, blah…"

In order to communicate with the limbic system, we must speak a language that it understands. And, guess what? It's not words. This *Far Side* cartoon provides insight into the way our limbic brain works. Ginger might understand something about her master's tone of voice, but she knows little about what her master is yelling. Likewise, our limbic brain will not understand how to accept, adapt, and creates change if we try to reason with it in words. Letting-go coping techniques are the limbic brain's language!

> I learned in the mind/body group to think about my life through the filter of mind/body approaches. They connected me to myself and to what was happening. It made me feel like I was taking control of my fertility, even if I wasn't pregnant yet. I no longer felt like a victim. I continue to use these techniques today. *Brenda, stay-at-home mom, Santa Monica, CA*

## The Unconscious Mind and Our Resistance to Change

The second obstacle to making changes as freely and easily as we'd like to, has to do with the unconscious mind. This section picks up where I left off in chapter 3 and develops the concept in more detail.

The unconscious is what it sounds like it is—the part of our minds that knows more than we know that we know. It is a *vast*–and I really mean vast!—reservoir of untapped intelligence and potential.

See if this clarifies the concept of the unconscious mind: Research has revealed the enormous intelligence of one cell. Bruce Lipton, a cell biologist and the author of *The Biology of Belief*, says each cell is a "miniature human"[32] in that it "possesses the functional equivalent of our nervous system, digestive system, respiratory system, excretory system, endocrine system, muscle and skeletal systems, circulatory system, integument (skin), reproductive and even primitive immune systems."[33]

Let's say that you were to look through a microscope at a one-celled amoeba. The amoeba is swimming in a drop of water. If you were to touch the edge of the drop of water with the tip of a glass rod that you dipped into sulfuric acid, you would see the amoeba swim away from the noxious environmental stimulus. The amoeba has a mind/body awareness of its environment and it responds accordingly.

We are comprised of fifty trillion cells, *each one with the awareness of a single amoeba*. Lipton calls the evolution of human beings "the consequence of collective amoebic consciousness."[34] Our cells are conscious of each other. One way to describe the unconscious is to say that we are not conscious of our cells' consciousness of each other. When Kim ignored her stressful symptoms, she was choosing not to respond to the "sulfuric acid" in her life. Since awareness is the foundation on which we achieve change, the lack of it kept Kim figuratively swimming in "sulfuric acid" for two years.

Awareness is about what we know that we know. Our thoughts are electrical impulses or energy. Our body is matter. Electrical impulses (energy) "communicate" with our biology (matter) by converting the electrical energy of our thoughts into neuropeptides known as hormones and messenger molecules. The research of Candace Pert, PhD, chronicled in her groundbreaking book, *Molecules of Emotion*, provided scientists with a new paradigm in 1998 that proved the vast communication capacities of our cells with one another. She states it this way, "The mind resides in every cell of our body."[35] This is the mind/body connection in a nutshell. What we think has impact on who we are. And by way of feedback loops, who we are influences what we think. To know this, is to live with awareness of the mind/body unity. *What allows for healing is the mind and body's awareness of each other.*

Here's where the plot thickens. How can we change thoughts and behaviors if they are out of conscious awareness? For one thing, letting-go skills allow us, by releasing the spasm of stress, to move toward making the unconscious, conscious. Let's take the little bit of science described above and elaborate further.

Our conscious mind, a function of the neocortex, is in competition with two more primitive structures—*primitive in terms of an earlier evolution, not primitive in terms of sophistication.* These two structures, the brain stem and the limbic system, respond to stimuli more rapidly than our neocortex (thinking brain). Early imprints form in the limbic system and the brain stem and result in the establishment of our underlying beliefs. These beliefs take up residence in our unconscious mind and control our thoughts and behaviors *before our cognitive mind has a chance to.*

Furthermore, humans have a propensity to scan for the negative. If cave man had not been wired for this, the saber tooth tiger would have rendered him extinct, and we wouldn't be here. Infertility has a talent for magnifying our in-born inclination to fix on the negative. However, if we decide to use our conscious *intent* to tame automatic negative thinking, we can override the oppressive demands of that negative thinking. You learned cognitive restructuring at the end of the last chapter. When practiced, it can move you along a continuum, so that over time you can become less and less at the mercy of negativity when it comes to making important changes. In other words, by bringing awareness to your bodymind and your behaviors, you can intervene in the mind's unconscious processes.

Lack of awareness is an unconscious phenomenon and is so powerful that our unconscious beliefs rule as if they were genetic. On the one hand, you can't be held responsible *for* your unconscious thoughts and behaviors. On the other hand, you can hold yourself responsible *to* them by building awareness through self-study.

Our unconscious mind makes it difficult, but not impossible, to set our intentions to change our style of coping. Simply by living life, the unconscious mind receives its programming. Mind/body techniques give you a chance to negotiate with the agenda of the unconscious mind and free your conscious mind to follow its conscious intentions with congruency. Thanks to neuroplasticity and neurogenesis, ultimately you can teach the unconscious mind a new set of rules. Mind/body awareness gives you a chance to be fully alive in your body and mind. Thus, you can heal the wounds that the infertility battle exaggerates. Chances are these wounds have been accompanying you through the years anyway. Think about this for a moment and you'll realize that infertility, though fraught with danger, really *is* an opportunity.

Chapter 7

# Hypnosis: The Ultimate Mind/Body Letting-Go Technique

In our fast-paced, multitasking world, doing is much easier than being. Yet, I've shown you the way to trump this inclination by understanding the forces of the limbic system and the unconscious mind. This chapter zooms in on hypnosis as a powerful intervention to the frenzy that tends to send people off searching for what to do without realizing that letting go into just being can be beneficial in so many ways.

Granted, problem-solving *doing* has its very important place. But letting-go techniques such as the relaxation response, listening to guided imagery tapes, focused breathing, various forms of meditation and hypnosis provide special advantages because they speak limbic language and they effectively break the spasm of mind/body agitation.

What letting-go techniques share in common is focus on the breath. The breath is a metronome that marks the presence of the present moment and brings the nourishment of oxygen to the body. Sometimes we forget that the breath, even one rich, deep breath, can be a tranquilizer. Take a deep breath now and remind yourself of this truth of nature.

Any of these techniques are the best way to return your body to neutrality and receptivity from the state of frenzy that the infertility experience has most likely been provoking. Engaging in these methods provides an inner oasis where relief and repair have the slow motion time to simmer and brew.

No doubt, you are well aware of the many vexing questions and decisions that require answers that do not seem accessible. It is inspirational to think about the way letting go opens up a space for the answers to make their way into consciousness.

Just the other day, a woman with whom I had worked hypnotically a while back, and who had conceived at age forty-four, came in to sort out her confusion after miscarrying. She wanted to get back in the game, but if she did, she worried that she could be pregnant when she was expected to be at a meeting in Europe. At the end of a very relaxing hypnotic interlude and said, "I've got it. Now I remember that after my miscarriage I thought, 'I really need a vacation.'" When the postmiscarriage upset moved to the center of the stage, she had forgotten about that. "I'm going home and calling my travel agent," she said. "What a relief to anticipate a week in the Caribbean with my husband. I'll wait until after my meeting in Europe to start again." This woman was able to create an eye in the storm and by doing so, recognized in trance what she "knew" all along.

While clearing a space for reflection by any method may have gotten this woman to the same place, hypnosis has one major advantage. It establishes a healing space where the unconscious mind can "be" receptive to posthypnotic suggestions. During relaxation, when the body has released physical tension and the mind has released frenzied mood, the limbic system can let go of the worry it has stored and be available to get beyond learned limitations. It is a powerful tool, indeed. You will witness this in the stories that follow.

## Entrance Yourself

If the state of mind/body agitation works against you in your quest to be emotionally and physically receptive, then the state of being in a hypnotic trance does exactly the opposite. Hypnosis creates an oasis—a very personal, soothing oasis in which your focus is held in the present moment. Body states can shift with the relaxation that accompanies hypnosis. Mental relief can come too, when the hypnotherapist guides you to a positive, safe, focused state—a daydream of sorts. The limbic system allows a respite from worries. In fact, hypnosis and infertility are both ultimate mind/body experiences; therefore, hypnosis can be the perfect antidote to the mental and physical stress of infertility. Posthypnotic suggestions, aimed at the unconscious mind, are what provide the benefit, making change possible.

Virtually all of the fertility patients with whom I have worked hypnotically have remarked on the enormous relief that hypnosis provided from the mind/body frenzy in which they felt trapped. I have used hypnosis with my mind/body groups, where I generalize posthypnotic suggestions to encompass all of the participants. I've also taught the group members self-hypnosis.

In the example of Julie that follows, you will see the power of hypnosis to heal emotional issues. Then, as you will see with Lori in the second example, you will understand what a boon hypnosis is when used to enhance aspects of in vitro fertilization. And you will be shown the way that hypnosis solidified the union between Angelo and Amelia in the context of donor insemination.

As you read the hypnosis scripts that go along with these stories, I believe that you will see that hypnosis is quite user friendly. The hypnotic subject is not under anyone's spell. The hypnotherapist treats the subject with utmost respect, unlike the case with stage hypnosis. Unfortunately, stage hypnosis has contaminated this approach, even though hypnosis has the approval of both the American Medical Association and the American Psychological Association. If you are someone who might be afraid of hypnosis, I hope this chapter changes your mind.

Before you read each of these scripts, take a generous cleansing breath as if to say to *your* mind and body, "I can let myself be immersed in these words." Perhaps this can prime you to have a deep and clear sense of the power of hypnosis and inspire you to learn self-hypnosis.

After you have gotten a sense of what hypnosis is like by reading and even indirectly "experiencing" these transcripts, I will deconstruct the concepts of hypnosis and prepare you to learn self-hypnosis.

## Reclaiming Femininity

Julie, a researcher at the time that I met her, had been anorexic as a teenager. She gave up the symptom when she left home to go to college, where she felt relieved and free to run her own life. Despite the fact that anorexia is a disorder in which the person banishes feelings of hunger as a way of establishing control, it wasn't until Julie literally found herself—to use the biblical term—barren, that she realized the ways that her version of control had backfired on her.

Perhaps, in some way, she had shifted the family pattern from feeling controlled to controlling the controllers by worrying them about her health. Now, years later, the abuse to her body was slapping her in the face at this crucial time. When she arrived at her red light in the quest for parenthood at age thirty-two, she had an imperfect sense of her femininity because she was not getting her period. She told me, "I realized that I had driven my body too hard for too long."

Julie's body was telling her it needed to heal. On the one hand, she felt that the "type A, control-freak part of [her] went along perfectly with the way that in vitro controls your cycle." This was both a boon and a bane to Julie, because the drugs gave her what she needed—ovulation and a buildup of the uterine lining—but on the other hand, she felt that the IVF process increased her sadness about being disconnected from her femininity. In addition, the

IVF process contributed to Julie's sadness because it "lacked the human element," despite her appreciation for the science of IVF.

Julie became aware after joining my mind/body group that "[she] had a lot of fear around experiencing [her] body doing what it was supposed to do." She also became conscious that her job and her lifestyle were "insane" and that she did not have enough time "to let [her] body relax." She told me, "I became aware that I wanted to get more grounded and healthier." The stress of struggling with her inability to conceive was so debilitating that she had inadvertently lost weight, which left Julie feeling that she was not totally on the other side of the anorexia. She wanted to be healed fully. As is sometimes the case, to be fully healed mandates scrutinizing lifestyle issues.

Julie conceived, but as I mentioned in chapter 4, she miscarried at sixteen weeks. But by this time, participation in the mind/body support group, acupuncture, and a commitment to a fertility-focused qi gong class had helped her to become aware of "just how vacant the ovary area of [her] body was and how much fear there was for [her]." Julie and I worked hypnotically with this issue, as you will see. Hypnosis connects mind and body. Julie's limbic system relaxed in trance, and her unconscious mind could take on reminders of Julie's femininity and its capacity to menstruate on its own. She began to feel in full flower as a woman.

Julie summed up what she became aware of that needed to change. Her words explain how she felt the hypnosis connected her mind and body in the way that she needed it to, with heartening results:

"Relieving the stress of being overworked, by quitting [my] job and freelancing instead, had allowed [me] to have more balance in my life. [I] gained some weight and felt healthier. The qi gong, acupuncture, and hypnosis were instrumental," she told me, "in building awareness of the mind and body connection and reducing the fear." In addition, she said "the support of the group and sharing the burden with trusted others helped me to get my body into a much stronger state of mind." Isn't that an interesting phrase that she used: "getting my body into a much stronger state of mind"?

Julie arrived at a place of greater tranquility. She had come to feel that grieving the miscarriage, and the expansion of her self-awareness, allowed her to be more at peace with the way her body functions. She felt ready to try to conceive again and she did, with twins! She had been afraid of her body processes, but she reported that it had been "extremely important" to work with me hypnotically during the IVF process and through the second pregnancy to visualize inside herself. She said, "You helped me to see two strong, healthy embryos, their vibrancy, and my body supporting them. It was hard for me to see that at first, but I came to feel that if I wanted it to happen, it would. I was able to remember and utilize my Grandma's Irish expression: Think a good, come a good."

Julie came to share my belief that her body could change as her mind and body became "friends" with one another again. After her twins were born, spontaneous periods returned. Furthermore, Julie used her grandmother's folklore, "think a good, come a good," as a positive affirmation to squeeze out negative thoughts. It may have been part of her grandma's folklore, but it has scientific merit as an example of Candace Pert's discovery that the mind resides in every cell of the body.

Furthermore, the mind's ability to influence the body is an example of the placebo effect, which in some studies is as high as seventy percent. What mechanism operates when people's health improves if they take an inactive drug? Herbert Benson's work on what he calls *Remembered Wellness* has shown the importance of the belief system of the patient and the doctor and the relationship between the two.[36] Julie and I had a strong connection, and both of us believed that to "think a good" *would* "come a good." Julie was working with a high quality reproductive endocrinologist, which matters, too. However, her individualized hypnotic intervention was a specific and targeted force in the armamentarium against stagnation in the mental and physical quicksand of infertility.

## Zooming in with Hypnosis

When in a hypnotic trance, the bodymind is encouraged to become receptive to ideas that the hypnotherapist presents as posthypnotic suggestions. Ideally, trance invites the workings of the conscious mind to tolerate stepping back as an observer. This enables the unconscious mind to come to the forefront, free to scrutinize and accept new or long-forgotten life affirming ideas.

In hypnosis, Julie could reconnect with her femininity and prompt a cascade of mind and body chemistry that contributed to the spontaneous return of her menstrual cycle. Furthermore, her grandma's expression, Think a good, come a good, was a comforting and powerful dose of love from her grandma, who was deceased, but alive in her heart.

Julie had come to believe and trust in her ability to conceive again, connect with her babies, and carry her pregnancy to term. The following is an excerpt from a hypnosis session in which we worked with these goals.

As is my custom, I taped this session for Julie, and she was enthusiastically compliant about listening to it regularly after our session so the she would have the best chance to integrate the posthypnotic suggestions for change into her outlook. As you read these words, imagine how well matched Julie's goals were with the ideas and the language of the scripts. Let yourself wonder what *you* would need to hear while in a relaxing trance that establishes a communication system between *your* mind and *your* body. What you need to hear is what you can establish as your goals. Make note of your ideas so you can use them when you learn self-hypnosis at the end of this chapter.

## Julie's Trance

*Give yourself a gift right now, Julie, of a nice, long, slow, and rich inhalation. . .allowing it to establish a pattern of breathing, which continues to bring solace and harmony to you. That's right. Allow your attention to turn inward, making whatever adjustments you need to for your best comfort, and legitimizing letting go of outward awareness. . .awareness of the sounds of traffic, of an itchy nose, or of anything that might be a disturbance. . . . Feel yourself sinking down, maybe in slow motion. . .the way you might feel if you were being enveloped in a trusting and loving way into a soft, inviting feather bed. . .sinking in slow motion into your own interior. Let it be a feeling of peacefulness, comfort, and safety. And experience the way you can use your breath to deepen the awareness of turning your attention inward, as your chest gently rises and falls.*

*You can enjoy arriving fully inward at the core of your being, by picturing yourself going down a lovely staircase—a staircase of marble or polished wood or one that is plushly carpeted. Perhaps the staircase in this fantasy takes you from the top of a beautiful bluff down to where there is a sandy expanse that draws you into your own sensuous pleasure—coming down the stairs one at a time, anticipating the warmth of the sand and the feel of it squeezing between your toes, already hearing the sound of the waves crashing and swushing onto the shore. You can feel the sun on your body, and you can smell the salt air. And, as you step down onto the sand and walk toward the water, feeling in touch with nature without. . .and within you, knowing that the ocean spray can cool you, and the gulls can delight you, enhancing the depth of your pleasure.*

*And now Julie, using the power of your mind's eye, provide yourself with a place to rest, perhaps a lounge chair or blanket, and allow yourself to sink down as deeply as you are able, feeling peaceful and supported. Spend a moment to allow the strength of the sun to literally bake you, taking all of the fight out of your muscles so you can feel limp and totally relaxed. Now give yourself another gift, Julie, of another deep and luxurious breath, feeling how something like the air that we breathe can feel just as satisfying as food when we are hungry or water when we are thirsty. Allow yourself to feel your lungs expanding with gratitude and relief and joy. . . that's right.*

*And, Julie, I can remember a time when I was a freshman in college, and for some strange reason I remember that it was October twelfth, and it was a glorious Indian summer day. I took a walk after biology class to the far end of the athletic field, where there was a low brick wall. A lot of times, students went and sat there with their backs to the field because if you looked out, you could see the Raritan River. I felt a propensity to take myself there just to sit and be at peace watching the river, and along the way, because it was such a glorious day, I remember lying down on the wall, soaking up the rays of the sun as you might be doing now on your lounge chair in your mind's eye. I felt that feeling of being baked by the sun, and I felt whatever tension was in my muscles just melt away. It was quite a dramatic feeling, and it has stayed with me all of these years. But, the important thing was that as an eighteen-year-old, I had my own struggles, my own tensions, and my own experiences of nervousness that resulted for me in unpredictable menstrual periods, just like you. But, I gave myself this time to merge with nature, and to my delight, I got back to my dorm room to discover that I had gotten my period. So allow yourself, Julie, to be in your comfortable place, soaking up the rays of the sun, and with it the sun's power to transform, to feel that all is right with the world right now, and just being here and at peace is all that matters. And your body knows that.*

*Your body has already produced a spontaneous period for you, and how nice that you can trust that your body is cleansing itself, that the endometrial lining would not be shedding itself if it hadn't built up just perfectly. Your ovaries were responding to your pituitary. . .creating a follicle. That follicle knows exactly how to mature, to burst forth, and then that follicle becomes its own little hormone system. . .releasing its chemistry, telling the estrogen when it should rise and when it should fall, telling the progesterone when it should rise and when it should fall. And as your uterus continues to shed and cleanse itself and prepare for the next cycle, your pituitary will once again release follicular stimulating hormone, which will have its impact on your ovaries, raising the level of estrogen in your blood stream, creating a feedback loop which brings the follicular stimulating hormone down, leutinizing hormone up, bringing you to your next ovulation and once again as that follicle bursts, your ova is released and drawn into the fallopian tube while progesterone is released to support gestation.*

*It seems now, Julie, you can once again trust your body. Wouldn't it be wonderful if your pregnancy and the unfortunate pain of its loss really served a very important function—to reestablish your femininity and your womanhood. . .to reset the dials, showing you that your body has remembered what its job is and how to do it. Take a few breaths, Julie, and use them to reinforce what your body has shown you that it knows it can do. Experience it with gratitude and loving acceptance, with receptivity, with an open heart, with joy and with an inner smile. That's right. Just like that.*

*While you are at it, allow yourself to take another deep breath, one of congratulations for a job well done. You can take credit for your participation in reclaiming your body. You deserve to honor yourself in this way. So take a moment to positively affirm the part you played in your own healing—thinking a good and coming a good. That's right.*

*And now take a while—it might be a short while, but it might feel like a long while—to allow your unconscious mind to do whatever it needs to do to support you, to help you to validate your body/mind connection. And, find within yourself the belief in your own powers. That's right. . .just like that.*

*And now with as much tenderness as you deserve, very slowly and gradually begin to shift your conscious awareness from the beach where you have taken yourself to enjoy your womanhood, and bring yourself back to the reality of this room, content in the knowledge that you can go back to this place, daily or even twice daily. . .building confidence each time, and then, returning to the reality of this room with a sense of health and harmony and true accomplishment, and whenever you are ready, open your eyes.*

As our work went on, Julie built trust and confidence in her ability to live in a female body that was not barren. She became more and more connected with her internal workings and we were, to a large degree, able to dispel her fear of losing the next pregnancy. After all, we cannot "*un*have" past history, but it is not a guaranteed predictor of the future. She could take inspiration from the pregnancy that she had held for sixteen weeks and was able to "think a good" about holding the next pregnancy in her womb to full term.

I spoke to Julie in subsequent trances about metabolizing the sadness of the miscarriage to make room for optimism, enthusiasm, and success. I reminded her of her strength and endurance. I reminded her of her growing capacity to acknowledge and then dismiss the negative thoughts that are a natural part of the human condition, in general, but of infertility for sure. I suggested a natural segue of learning to care for herself—love herself, if you will—to caring for and loving a baby, making that real. I congratulated her for being able to question and soften her fear and replace it with love for and from her husband and others who had been so supportive once she opened up to them. The love that her precious grandma had imbued in her was prominent.

By the time Julie arrived at IVF, which would give her twin daughters, she was well-fortified with a strong sense of herself, a sense that had gone missing with her appetite in the time of anorexia. Julie not only reclaimed her womanhood, she reclaimed the sense of herself that she had before the arduous journey toward parenthood became an issue. She has come to feel better than new.

## A Little Bit More Science

Hypnosis is a boon as a mind/body tool, as well as a tonic for the whole process of in vitro fertilization. Clinics do their best to run efficiently. In general, what is most needed at this time, the human touch, often feels overpowered by technology and the business of IVF.

Anyway, who in their right mind would volunteer to show up for blood work or a sonogram at the crack of dawn? The atmosphere of even the best-run IVF clinics, in combination with the demands of evaluations or treatments, results in a situation that is the opposite of what a mind/body doctor would order.

IVF is expensive and an ordeal. Women feel that so much rests on this or whatever procedure they are going through. Stress increases if this is the last or only IVF the couple can afford. Ovum retrieval and embryo transfer are critical times and difficult to endure. For that matter, so are the drug protocols that launch the process. They increase emotional vicissitudes, making the experience of IVF feel like PMS on steroids. The two-week-wait to find out if you are pregnant is universally reported to be the most difficult part of it all.

Hypnosis, as I have said, can be the perfect tonic. The hypnotic subject is encouraging their body to work cooperatively with the chemistry of IVF. Mental participation while in trance can be used to "rehearse" for an optimum retrieval and *given the workings of the limbic system, which does not know that you are not where you're imagining yourself to be,* rehearsing is an important thing to *do* and an important place to *be*. If nothing else, it is a stopgap measure to head off preoccupation with negative thoughts before they convert to neuropeptides, which send the wrong message to crucial receptor sites.

While it is true that we cannot prove that hypnosis contributes to or facilitates drug efficacy, we know from other studies that setting mental intention while in trance affects the body greatly. To name just one study by Carol Ginandes, PhD, subjects with a certain kind of ankle fracture were divided into two groups. The control group was casted and sent home. The experimental group was casted, but also learned how to use self-hypnosis to "ice" the injury and send healing intention to the ankle. The experimental group healed weeks ahead of the control group.[37]

There is no reason to believe that the same could not be true with an intention to boost reproductive functioning with hypnosis. Conversely, there is science that shows that thinking a bad can come a bad. Herbert Benson, MD, calls this the "Nocebo effect" in his book *Timeless Healing.*[38] The most dramatic example of this is how in certain indigenous cultures, if a "death spell" is cast on someone, they've been known to become so terrified that they actually die. We can witness the nocebo effect in our modern world. Perhaps you know someone who dwells on a worst-case scenario, and the next thing you know, their bodies have followed the path of their minds.

Here I must add a disclaimer of sorts. While thinking positively or at least curtailing thinking negatively, spares us the agony of being stuck in a black mood, and has been shown to matter as it did in the study about letting-go coping, it is important to recognize that there are unknowable factors that figure in, including plain luck. *The last thing that you need is to blame yourself for not getting pregnant because you thought the wrong thing or failed to achieve relaxation at a particularly stressful time.*

That notwithstanding, hypnosis does appear to be more than a tonic around embryo transfer. I encourage my patients to come in for a hypnotic session before transfer based on the study out of Israel, reported in the May 2006 issue of *Fertility and Sterility*, which I mentioned in chapter I. Remember, this study divided a group of IVF patients in half. The control group went through IVF as usual. There was one difference with the experimental group. At the time of embryo transfer, this group received a hypnotic intervention.

During transfer, of course, the uterus is opened. When it is, it is likely to spasm. The experimental group received a posthypnotic suggestion in trance to extend the bodily relaxation to encompass their uterus so they could keep it as still as possible despite it being opened. It is felt the acceptance of this suggestion might explain why the pregnancy rates of the experimental group was *double* that of the controls. It was hypothesized that if the spasm of the uterus was minimized, the beautiful embryos weren't being "spit out."

I make personalized hypnotic audiotapes for my patients in anticipation of embryo transfer, so they can build competency in relaxing into the transfer experience. Patients feel as if they are participating in an important way, mitigating the feeling of being out of control.

## Healing Three Issues with Two Trances

Lori is an assistant to a top executive in a financial firm. In her individual work with me, Lori looked forward to using of hypnosis to heal the emotional wounds that came to awareness as we worked together. In addition, she enjoyed the connection she had come to feel to her body. She had worked with me in this way for a year before she was ready to go back to the reproductive endocrinologist for the IVF that resulted in her pregnancy.

Before her first appointment with the reproductive endocrinologist, and a few months before I met her, Lori learned by way of surgery that her fertility quest dead-ended because of asymptomatic endometriosis, an ovarian cyst, and a uterine polyp. In addition, she learned that she had a tipped uterus and that her tubes were blocked. She and her husband realized that they had "wasted two years trying on their own." When I met her, she was in a dilemma—the clinic was ready for her but she was not ready for the clinic.

After her surgery, Lori confessed that she was "terrified" of in vitro fertilization because of a fear of the drugs that are designed to induce superovulation.[39] She had had an experience earlier in her life in which she had an idiosyncratic (unusual but particular to her) reaction to an antidepressant drug. Instead of relieving her depression, it intensified it and created a terrible anxiety as well. She had gone into a serious emotional tailspin at the time and feared that any drug intervention now might produce the same result.

In the year that we worked together, she needed to exhaust the possibility of conceiving with a holistic approach. This was a good plan because it gave her time to experience low doses of the ovulation-inducing drugs and brought her to a place where she understood that her history would not repeat itself. The holistic approach did not work for her. IVF was the next logical step, and a step she could now take. She had overcome her fear of the drugs. We also worked with hypnosis to allay her needle phobia, since injections are a mandatory part of in vitro fertilization.

In that year, Lori's self-awareness grew, and she was able to simmer down her inclination toward being high strung. She took excellent care of herself, went to yoga class regularly and got a massage as needed. She was feeling empowered by her capacity to find "the eye and the I in the infertility storm" with the mind/body techniques that I had taught her, and she loved the hypnotic work we did. Lori established appropriate boundaries between herself and a boss who was what she called "a type-A maniac." She was pleased, too, to feel as if she had separated herself from her family's emotional system where "everyone was a nervous wreck and depressed." She felt mastery over pessimism and negativity, and told me, "Even when I feel these things, I know it's temporary." She had done a fine job of clearing the path for IVF.

As is sometimes the case, setbacks and disappointments set in. Her husband was diagnosed with a morphology problem and had a varicocoelectomy.[40] She developed another intrauterine polyp, which required surgery. Her husband's father, from whom he had been estranged, died. This created an emotional spin in her husband, which added extra emotional pressure on her. On top of this, Lori's company was bought up and in the merging process, demands on her intensified. There was much to contend with, but Lori was heartened at how her decision to discipline herself by practicing mind/body techniques kept her fortified.

With the first IVF attempt, the flattening of the menstrual cycle with Lupron was what she called "too harsh," and her doctor cancelled the cycle. Then she discovered that before she could schedule the next IVF, she needed surgery again for yet another ovarian cyst and more endometriosis. In Lori's words, "I needed four surgeries just to get to the starting gate."

Riding the waves of anticipation and disappointment was not easy until the time when finally "all systems were go." During this time, Lori depended heavily on our hypnotic work to get her through the next leg of the challenge. She was determined to remain open and receptive. She said, "I understand that thoughts have wings, and I want to keep my thoughts positive."

I made an audiotape for Lori before both retrieval and embryo transfer. Both of the scripts follow. You will learn how hypnosis can feel like a safe haven at a highly charged time and how you can foster the connection of mind to body when it is most needed.

Once again, I recommend that *you* take a cleansing breath right now and bring yourself into the present moment with Lori.

## Lori's Retrieval Trance

*Take a nice breath Lori, and experience your life force even as you prepare to create new life. Use your breath as if it were a broom, collecting anything that you no longer need and sweeping it away with the exhale. Feel the way that you can also use your breath to soften and loosen your muscles and even your internal organs. And isn't it interesting that intuitively, you put your hands on your belly. I wonder what your hands might be saying to your ovaries? Perhaps they are saying,*

*"I know what flows through my blood stream is reaching its destination and is preparing to release my half of the genetic union between Ken and me. Thank you. I'm grateful."*

*And, you can inspire your creative imagination to wonder: what if each of your hands had one hundred little ears. What would they be hearing from your ovaries, which they are supporting and caressing right now? And, what if those hands had one hundred little mouths that could speak to those developing ovum? What would they say?*

*You know by now, Lori, that hypnosis is both a conscious and unconscious process. And you can rest. . .assured. . .that your unconscious mind is hearing and speaking—communicating—sending and receiving love and encouragement to your ovaries in just the right way.*

*And you can allow the tumult of the clinic to float away—the tumult of getting there and the hubbub of being there—can float from your awareness like a helium balloon. . .up, up, and away, farther out of sight and out of mind. Meanwhile, you can bring yourself down into your body. . .pleasantly down, arriving at the center of your being where your truth and wisdom reside.*

*And, from this place of safety and comfort you can bring even greater relaxation to yourself, visualizing a favorite place in nature, a place that you've been before, or one that you create for your own delight. And I don't know, and I don't have to know, whether you are enjoying a lake. . .a waterfall. . .an ocean vista. . .a meadow. . .or a garden. Just allow yourself to explore the way that your body can feel so soothed by nature's bounty, whether it is the scent of your favorite flowers, or salt air, or the way that your eyes can feast on the multitude of colors, the way that your body can be warmed by the sun or cooled by the breeze, the way that your ears can transport you by way of the rustling of the leaves in the wind, or the sound of the birds singing to one another. And wherever it is that you take yourself is exactly right. Feel the way that your body has come down another notch, one breath at a time.*

*And now notice, Lori, how right in front of you is a perfect place for you to rest. Perhaps it is an upholstered chair. Perhaps it is a hammock. Perhaps it is a soft blanket on the grass that allows you to look up at the sky and let your imagination run wild as the clouds travel and change shape.*

*Settling in, Lori, bring your attention back to your hands. Feel their warmth on each other, their warmth on your belly, and your belly's warmth on them. And did you know that the hands, the lips, the tongue, the ears—all of the parts of us that have to do with communicating and being connected—have the most neuronal connections in our brains? So your instinct to put your hands on your belly was excellent. This is about connecting and communicating with yourself. . .and with your future.*

*And once again, Lori, trust the power of your unconscious mind to make sure that those messenger molecules that come from your heart—that are longing for the time that will come when you will meet your baby—those messenger molecules can travel from your heart to what is exactly under your hands. The messenger molecules can support the medication that is coursing through your bloodstream right now, making sure that your uterus will be a perfect nest.*

*That's right, trust your body, Lori. Your body is responding as it is supposed to...and you can allow yourself to pace forward, one day at a time, as your ovum continue to mature, continue to stretch toward your hands like little birdies opening their mouths to be fed; they're coming forward to the surface of your ovaries—all the easier to be retrieved.*

*The expert care with which they will be handled—and the gentle way that your genetic material will be put together with Ken's—take a moment to trust that Lori...trust it!*

*And the energy of those genetic cells, the greeting they give each other, the way they know exactly what to do, the way those embryos will form and will instantly begin to respond to their imperative to develop, growing...growing exactly right. Trust your medical team to know when it is the perfect time to reintroduce them back to their home inside of you.*

*Take a moment to feel your receptivity. Breathe into it. Make it real. You've done everything in anticipation of this time and you've done it well. Trust the way the door of receptivity will be opening for you...very soon.*

*And, before you prepare to return to the reality of this room, connect with your hands and their power to connect. Be still with them so you can memorize the flow of love and energy and communication. Return to awareness of your breathing, the life force, and its implications and as soon as you are ready...open your eyes.*

## Lori's Embryo Transfer Trance

Lori was compliant in practicing with the retrieval tape. The retrieval went well and she was most interested in giving herself time to practice keeping her uterus as still as possible during the transfer. We made another tape. She had been feeling optimistic and had gone into Pottery Barn Kids. It was an act of faith to allow herself to go there. In the process, she had fallen in love with simple white furniture and bedding that was a soft blue with puffy white clouds. She allowed herself to imagine that this would be the décor of her child's room. I used her imagery in the trance, as you will see.

You may think it strange or risky to go so far as to imagine that in vitro fertilization will be successful without having a guarantee. Most people, if left to their own devices, shy away from this, fearing that the disappointment, should it not work, would be infinitely more devastating. Truth be told, the disappointment will be the same, and it will be painful. You cannot protect yourself from this kind of disappointment, so you may as well do the best you can to believe in a positive outcome.

Meanwhile, Lori's version of Julie's grandma's old Irish saying Think a good, come a good, was thoughts have wings" She was consciously seeking to live in the unity of her mind and body as she took herself into Pottery Barn, thinking positive thought.

I did not tell Lori to go to Pottery Barn, but I utilized it and her hard-earned positive attitude by engaging her in the imagery of her baby's nursery. Once again, I invite you to imagine the impact that *your* imagery could have on *you* as you read these words. Make note of what you would need to hear when you learn how to make a self-hypnosis audiotape.

*Take a generous inhalation, Lori, a big sigh of relief, as a way of congratulating yourself for the enormous victory that you've just told me about. How wonderful that you have reinvented yourself as someone who can trust the medical world, can appreciate the sophistication of how far these procedures have come. Now you are someone who can let go of needing to know all kinds of details as if that put you in control of them, when in reality you are relaxed about giving control over to the people who really have it—your medical team. What you can be in control of is in you, with every breath you take.*

*And the idea that letting go has calmed you down, has softened your whole experience as if you are floating into this process, now. You are not only floating into it, but expecting that it will work. What a gift you are giving yourself! So take a moment to feel how your body feels when you expect a positive outcome. This may be a conscious process—you may be thinking things that directly have to do with this kind of trust and happy anticipation. Or, this may be an unconscious process that goes on in spite of what your conscious mind is thinking. It will certainly be a physical process in which you can continue to let go and let down. . . letting those lovely, graceful hands of yours that are now once again positioned over your ovaries and uterus making their happy connection. And, just like that ad for the Yellow Pages, you can let your fingers do the walking, do the sensing, do the feeling, and do the loving. Your loving heart can come forward and connect with your womb as your busy brain slowly but surely stops jumping like a monkey.*

*And so as you continue to breathe freely and easily, feeling the rise and fall of your chest, you can allow each breath to take you deeper and deeper into your own victory, deeper and deeper into your own trust, into your own happy anticipation of the future and into a place of growing safety, security, relaxation—that's right. You can enjoy to the fullest the calmness and the tranquility that comes from this kind of victory.*

*And so, Lori, in anticipation of your transfer, transfer of your child or children, I'd like you to use the power of your creative imagination to take yourself into that pale blue room, with the puffy white clouds, and put yourself onto one of those clouds. Just allow your body to be received by one of those clouds as if it were a featherbed—strong and secure enough to support you. Allow your body, as it sinks gracefully into that featherbedlike cloud, to remind you of the rewards now of that loving heart and the way it has taken you to this very place of victory.*

*Notice the feeling of floating and being supported simultaneously—so comfortable, so secure. . .and for you, such a happy image. Just enjoy the feeling, letting your body relax even more. . .that's right. I once went parasailing so I've been up near those clouds, Lori. When parasailing, the stillness is exquisite. The vista is breathtaking. The sense of serenity is extraordinary. And, as you hear these words and as you imagine yourself being supported by one of your very own clouds, of your very own creation, in your very own house, in your very own baby's room, allow the feeling of support and stillness to take you deeper down into a place that your unconscious mind knows, a place to which your unconscious mind can guide you so that at the time of transfer you can have and hold that extraordinary stillness—stillness of your forehead and scalp, stillness of your neck and shoulders, stillness of your face. . .your arms. . . your torso. . .your legs and feet—but most important, stillness of your uterus.*

*Your doctor will need to open the door for those embryos to be transferred to their home—a home that will feel to them just like those clouds, mushy, soft, and receptive, yet strong and supportive. In order to do your best to encourage implantation, you can begin now to practice transferring the stillness that you feel all over your body to your uterus, so that even as it must be opened ever so slightly, it can remain as quiet as possible. The best that you can do is all that you will need. No matter what, Lori your body will figure out how to keep your uterus still in exactly the right way. And, as you practice and continue to build confidence in your power to keep your uterus still, it is as if you are saying, "Come to Mommy. I'm here. I'm ready for you. I can't wait to hold you in my body so that I can hold you in my arms."*

*And you can trust that this process of practicing stillness will go on all by itself, because from this oasis of rest and relaxation, the smartest part of you is alert and listening and cooperating. That's right.*

*And take a moment now to breathe a sigh of relief, knowing that the same way that you let go and let down the other day is just enough and just right, and that in the process of letting go and letting down, relaxing in the safety and support of the softness of the clouds, you have everything that you need to maintain utter welcome and receptivity to your embryos.*

*And as you feel your fingers moving, giving yourself a thumbs up, you can rest. . . assured that you've taken all of this in, and that you can continue to dream about the future and trust that dreams can come true.*

*Feeling that deep inhalation, acknowledging your truth, continue to experience your receptivity, your softness, the heart-felt longing and the love that has brought you this far. Allow that to ring true. . .allow your victory to ring true.*

*Take a moment now to memorize the body sensations of floating on your cloud, reminding yourself that you know how to get back to this place for practice and enjoyment. You have reason to trust that you will return to this place as the transfer is happening. And, in fact, you can trust that you'll be able to maintain this state for the rest of that day and the next day and you will know when you are ready to resume normal activity.*

*And now, Lori, I'd like you to prepare, slowly and gently, to shift your level of conscious awareness away from that delicious little room with those appealing clouds—for now—bringing yourself back to ordinary reality, confident of your capacity to go back into that safe space, confident that your unconscious is an ally, helping you to get where you need to be and stay where you need to be, practicing this for the next few days. . .and then whenever you are ready, open your eyes.*

Remember that the design of these two trances were meant to heal old emotional wounds by keeping Lori connected to her body in addition to facilitating the retrieval and transfer. Mission accomplished.

Lori's son was born in April of 2008 as a result of this IVF attempt. As is sometimes the case, a daughter came soon thereafter, as what Kim from chapter 5 called "a freebie."

## Hypnosis for Couples

Chapter 3 featured the journey of Angelo and Amelia. What follows is a transcript from hypnotic work with both of them simultaneously. You will see in this next excerpt that it is possible to humanize the technology and really experience the power of love to lead this hi-tech parade. As usual, you might want to imagine the impact that this trance could have on you and your partner.

*Take each other's hands and now take a deep breath and allow your eyes to softly close... that's right. And, as you exhale, feel the way that an exhale has a movement. Feel the way it seems to take you down. On the next exhale let yourself go down even more, into the center of your being. Just settle in as if you're coming in for a landing with a parachute...settle into who you really are...right into the middle of yourselves.*

*As you just rest in the rhythm of your breathing, allow yourselves to get comfortable within you, noticing your capacity to bring peace and tranquility to your forehead and scalp, smoothing out any wrinkles. Feel the way you can balance your head by bring your ears equidistant from your shoulders. Some people like to experiment with their breath. What does it feel like to inhale fully? What gets released on the exhale besides the breath? By now, your arms and hands, legs and feet, may be feeling deliciously heavy—resting and comfortably supported by the sofa. Take in another breath and use it to realize just how much power you have had to bring relaxation and comfort to yourselves in these last few moments.*

*Take a moment...now...to extend your awareness...so that it goes from the middle of yourself...and encompasses the middle of each other. And, you can...rest...assured that you are in exactly the place that you need to be—a place of unity, a place of purity, a place of truth...a place where you can recognize that your inordinate strength has served you very well. Take a moment to congratulate yourselves and each other for having all the strength that you've needed to get yourselves to this place...and to know that from this core of your individual beings you can feel the unity with each other...each of you honoring that you are in this together.*

*The energy of that unity can be realized...utilized...so that as you go through this process—as you surrender to the technology—you can still have what you have at this moment, surrendering to feeling your love for one another. Take in a generous breath, and on the exhale, experience the synergy...and energy of your teamwork. Let this supersede the idea that you've needed assistance along the way. After all, would there be flowers, were it not for assistance of the bee? And, so just feel the way that you can softly accept the help, soften and warm to the technology...with gratitude. Soften and warm to the synergy and energy of your partnership.*

*And, Amelia and Angelo, use this time to feel the warmth of each other's hands and the warmth emanating from each of your hearts...the warmth that really represents your authentic selves. And just sit with that for a moment, and enjoy it.*

*Enjoy it, and energize it, and more than anything else, trust that this experience will go with you through these next days and weeks. That any time that you greet each other, say hello, kiss goodbye, grab each other's hand, have a cup of tea together...the frenzy of your lives melts away, and your mind and body are reminded of this moment. Now take another moment or two to trust the teamwork of your conscious and unconscious minds, to solidify whatever it is that feels exactly right. . . whatever it is that will allow you to take this experience with you, and to trust it. Now, taking your time, whenever it is that you're ready, open your eyes. . . .*

This was the trance work that preceded Amelia's conception with their son.

## The Concepts of Hypnosis

These stories give you a sense of just how powerful hypnosis is as a mind/body intervention for individuals or couples. Trance is one of our natural capacities that most of the time we *avoid* so as to keep ourselves busy *doing* our lives. Trance is nothing more than the purposeful narrowing of focus with a willingness to ignore the "noise" of the conscious mind.

Infertility's demands ramp up the *doing*; hypnosis brings balance back when you choose to utilize it to let yourself *be*. Letting yourself be relaxing into a hypnotic trance, then becomes the backdrop for delivering post-hypnotic suggestions, which is where the powerful bonus of hypnosis comes in.

Since *you participate* in connecting your mind and body with conscious awareness, *you experience control* of helping yourself through the infertility journey, a process that reverses the feeling of being out of control. With IVF, you, in fact, must hand over control of your body to your medical team. Your body responds when and how the drugs tell it to. Jenn captured this necessary process with these words: "I feel as if my body has been hijacked." Hypnosis allows you to balance the necessity of allowing IVF to control of your body with experiencing connection to your body as you actively participate in the hypnotic process. Julie, Lori, and Angelo and Amelia became entranced within their own internal processes. This process is a salve that can heal wounds.

Well-trained clinical hypnotherapists understand the multifaceted, multilayered intricacies of hypnosis. I want to make self-hypnosis available to you as a tool. That means that I need, to a large extent, to simplify things so you are not so confused that learning self-hypnosis feels beyond your reach.

I'll start with deconstructing the above trances.

## Understanding Hypnosis

### *Concept 1: Establish the Goal*

For Julie, the goal was to reinforce and validate the process of healing the physical manifestation of her anorexia, which resulted in her losing her menstrual cycle. The healing process would involve addressing the fear of living in a female body by familiarizing herself with herself. That explains why I rehearsed the way the body prepares for menstruation with her.

With Lori, there were two goals. She needed to heal her fear of the drugs that are part of the IVF protocol. That fear was based on an earlier trauma in which a drug that was supposed to help, instead gave her a frightening experience. In addition, we were preparing her for mind/body participation in the IVF processes of retrieval and transfer.

For Angelo and Amelia, their trance validated their past efforts and suggested that their victory in redefining and living love in an accurate way had set the stage for the upcoming donor insemination to have a positive outcome.

***Concept 2:*** *Achieve Relaxation and Pay Attention to Relaxation Achieved*

Remember that the limbic system responds to messages that come from the brain stem (automatic functions such as breathing and heart rate) and the neocortex (pleasant imaginings). With these signals, the limbic system senses that all is right with the world and accepts the trance state. The body behaves as if it is where it imagines itself to be.

Therefore, as far as Julie's limbic system was concerned, easy, slow motion breathing encouraged the movement into trance. I ratified this by encouraging her to notice the rise and fall of her chest. She was thus set up to enjoy the delights of the beach. Likewise, the description of nature captivated Lori. This set the stage for communicating with her developing ovum and the formation of her embryos. In the transfer tape, she was absorbed in the comfort of the soft, supportive cloud where she felt nestled. Angelo and Amelia were invited to travel down into the center of themselves—a whimsical "place."

For Lori, focus on the breath, and involvement in the fantasy about nature in the preretrieval trance and the softness and support of the clouds in the trance before embryo transfer were the vehicles to achieving relaxation. She received feedback that she was becoming absorbed in trance with words such as "that's right" or "the warmth of your hands."

Angelo and Amelia were encouraged travel to the center of themselves, a whimsical place, and to melt into their newfound unity.

***Concept 3:*** *Deepen the Trance State*

I invited Julie to take herself *down* a staircase to the beach, deepening her absorption in the sensuous pleasures of body and mind.

I deepened Lori's trance by focusing her on her hands as instruments of connection and communication. In the transfer trance, I took Lori deeply into the safety of the cloud, which allowed me to take her beyond the victory of the retrieval and subsequent fertilization to imagining the victory of the transfer.

Angelo and Amelia were encouraged to notice the synergy between them and the transference of warmth through each other's hands, an absorbing, deepening suggestion.

***Concept 4**: Honor the Power of the Unconscious Mind*

In all cases, the concept of the unconscious mind is presented as a given, and an ally, that can be relied upon to utilize all of the concepts of hypnosis. I accomplished this with phrases like "You can count on your unconscious mind to support..." or "Your unconscious mind can solidify...." The relaxation and the deepening of trance set us up to suspend the conscious mind's tendency to nag and judge.

***Concept 5**: Post Hypnotic Suggestions*

This is where hypnosis gets even more powerful, given the relaxation and, hence, the receptivity that is established. For Julie, suggestions involved trusting her body and allaying her fears about her feminine functioning. I took her on a journey through her body's experience of the menstrual cycle as a reminder that she knows how to be a woman. Phrases such as "with an open heart" humanized the technology. When I said, "Affirm the part that you played in your own healing," I meant to call Julie's attention to her empowerment.

Both of Lori's trances set her up to be open to the expectation of victory of the retrieval and the transfer. This contained within it an inference (a subtle suggestion) about a victory in a pregnancy. I tucked suggestions in the first trance into the theme of connection and communication. In the embryo transfer trance, she could expand the stillness that she had achieved in her body to include stillness of her uterus. Since her imagery in this trance was of the baby's room where the baby would sleep and dream, I was able to suggest that "dreams can come true."

Angelo and Amelia benefited from the reframe of needing help ("Don't the flowers need the bees?"). And, they were prompted to "take this experience (of unity) with them."

***Concept 6:*** *Metaphors and Stories*

There is a childlike part of us that enjoys being in the state of just *being* in the present moment. As such, it enjoys stories and metaphors. In Julie's trance, I told her a parallel story about myself. I told Lori a story about parasailing. The imagery of the bee helping the flower to reproduce is right up the unconscious's alley.

Metaphors are a very important method of communicating with this childlike part of ourselves. The idea of the sun baking the fight out of Julie's muscles encouraged her body to give up the tension/fear about returning to full womanhood.

Lori's trances were loaded with metaphor. The breath as a life force was an analogue to creating new life. Words like "nest" and "home" were metaphoric for the uterus. The cloud, which took her above the fray of infertility to a place of just *being*, simultaneously took her to a place of being with her soon to be conceived baby. For Angelo and Amelia, the suggestion of the synergy and energy of their union was a whimsical metaphoric image of their melting together.

***Concept 7:*** *Anchor the Experience*

Since these trances were taped and meant to be listened to so that the messages could be reinforced, statements were made to insure that the suggestions would follow the person out of trance and remain with them, and also leave a trail that would be easy to follow back into trance the next time. Anchoring the experience means calling attention to this trance and suggesting that it will be easy to return to this state in the future. Therefore, I commented as each of them was coming out of trance: "You can remain content in the knowledge that you can go back to this place daily, or even twice daily," or "You can memorize the flow of love and energy and communication," or "You can remain confident in your unconscious as an ally, helping you to get where you need to be and stay where you need to be, practicing this for the next few days." To suggest "any time you greet each other" anchored a suggestion to Angelo and Amelia.

Now, with a willingness to put the effort into learning how to create a script, it can be your turn!

# Exercise

## Self-Hypnosis Simplified

There are many books about self-hypnosis. At the outset, you can't expect yourself to learn the myriad intricacies of hypnosis. If, however, you have familiarized yourself with the four trances and the deconstructed concepts in this chapter, you will have enough of a baseline to follow my guidance, write a script and then create an audio recording to which you can listen. If you want to develop your skill, you can refer to Brian Alman, PhD and Peter Lambrou, PhD's *Self-Hypnosis: The Complete Manual for Health and Self-Change.*

Although all hypnosis is actually self-hypnosis because no one can force you into trance, those of you who fear that you would be giving up control can feel reassured. Once you've made the audiotape, you will be listening to *your* voice expressing *your* goals, suggestions, and musings.

I'm going to provide you with a *very* simplified guide to making your own self-hypnosis tape. First you write the script. Then you tape it. Listening to the recording will then be a piece of cake because you won't have to memorize what you want to say to yourself. Self-hypnosis without a recording can come later when you feel more familiar with the process.

To proceed, buy or borrow an audio recorder. A digital recorder is great because it allows you to load your recording onto your computer in iTunes and then put it on an iPod so you can easily take it with you. You can purchase a Digital recorder for under $100.00. The only other thing you'll need is the willingness to create your script.

Before you can write your script, you must know your goal and your posthypnotic suggestions. Goals can be self-exploratory, stamina building, aimed at mastering a behavioral change, designed to reinforce commitment to a goal, or seek to strengthen an emotional state like self-confidence. Goals can pertain to minimizing emotional triggers or setting boundaries. To identify the goal, ask yourself What is the problem I want to work on? The solution becomes the goal.

Write your goal here:________________________________________

You must also identify your posthypnotic suggestion(s) before you start. These have to do with *visualizing yourself already at the goal state.* You can think of it as an imaginary rehearsal. The best way to break it down is to think of when "this" occurs I will do "that." Make your suggestions specific.

Write your posthypnotic suggestion(s) here:

________________________________________

________________________________________

________________________________________

## Writing Your Script

This fill-in-the-blank script follows the seven concepts deconstructed from the Julie, Lori and Angelo and Amelia's trances. Don't be afraid to elaborate on my model. Bring your own creativity to this script.

At the top of your script, in bold, put SPEAK…VERY…SLOWLY as a reminder to use hypnotic tone and tempo. In the frenzied world in which we live, it behooves us all to follow Lily Tomlin's advice: "For fast acting relief… try slowing down." "Hurry up and relax" just doesn't work if you want to go into trance.

1. GOAL: "Okay, (say your name), take a nice, slow, easy cleansing breath, as a way of separating *before* from *now*. (Pause) Feel the way that breath has started the process of letting go and letting down so that I can learn to (state your goal).

2. RELAXATION: "I can now… (choose one or more options from this list:)

a. … "focus on my breathing. I can really enjoy the feeling of expansion as an inhalation fills my lungs. (Pause) And I can enjoy the release of an exhale as I let go of as much of what I no longer need as possible." Elaborate, if you wish. AND/OR

b. … "bring my attention to my body. I can feel my power to smooth out my forehead and scalp. (Pause) I can soften my facial muscles. (Pause) I can let my lower jaw go slack and feel the cascade of relaxation that follows. (Pause) I can adjust my posture…and in the process, I can relax my shoulders away from my ears. (Pause) I can feel the strength of my spine as it supports me in an up…right posture. Now I can feel how easy it is to let my internal organs go loose and limp. (Pause) I can feel that delicious heaviness in my arms and legs. (Pause) I don't need them to do anything right now except to just *be*." Elaborate here, too, if you wish. AND/OR

c. ..."I image myself now at (a place–fill in the blank) where I can enjoy" (fill in the sights, sounds, smells, tastes, and feel of this setting. Be specific—what are you seeing, what are the colors? Do you hear birds? What can you say about their songs?, How does the aroma of something make you feel, etc.). "I can let these sensuous pleasures bring me to a place of ever-increasing comfort, relaxation, and safety."

3. DEEPEN TRANCE: "I can continue to enjoy this relaxed feeling by" ...go deeper into any of the above experiences and remark on how it is enhancing trance comfort. And/or you can literally take yourself down a staircase or elevator, for instance, as a metaphor for getting deeper into the experience. Make sure to elaborate on what you see every few steps or describe the physical sensations of going down in the elevator.

Take your time with steps two and three, all the better to savor the trance experience.

4. HONOR THE POWER OF THE UNCONSCIOUS MIND: We don't fall out of bed at night because there is a part of us that is alert and awake to this kind of reality. That's the part that I'm inviting you to trust. So, you can say, "I can trust that there is a part of me that knows how to (state the goal) and I can, now, ask that part of me to help me achieve (the goal)."

5. STATE YOUR POSTHYPNOTIC SUGGESTION HERE. Repeat it in different ways, with confidence. " I know that when I ___________, I will be able to _________", or "when __________, I *will* _________."

6. METAPHORS AND STORIES: Let your mind wander and trust that the unconscious part of your mind will find its way to something curiously related to your goal. "Now I wonder where my unconscious meanderings will take me?" Or "The childlike part of my unconscious mind will present me with a story or metaphor about accomplishing my goal." Or, "I can be pleasantly surprised as my unconscious mind presents me with a story or metaphor about my goal of ___________."(Leave some space on the audio tape here for the unconscious to respond. Tell yourself that you will be comfortable with the silence.)

7. ANCHOR THE EXPERIENCE in your memory and/or in your body: You can say, "As I bring this experience to a close, I can find a place in my body to store everything that has been important to me, making it that much easier to come back to this experience again." Or you can give yourself a "cue" by saying "I can return to trance, whenever it is right and appropriate,

by squeezing my thumb and index finger tightly together." You can create a transition back to ordinary reality in this section by returning to the images that you used to go into trance, by suggesting that you bring your attention back to your breathing and your body.

Keep in mind that you can make a tape for yourself and/or a tape for you and your partner, which would enhance your sense of togetherness and joint purpose. Once you've made your recording, you should clear a time and quiet, comfortable space in which to listen to it. Self-hypnosis is a skill and practice will build your confidence in having this as an important letting-go tool. You can identify new goals for finding the "eye and the "I" each time a new infertility storm blows by.

Before you actually listen to the tape, you might try one of these little tricks that can make the transition *into* trance easier. For instance, if you moisten your finger with saliva and wet the place in the middle of your forehead, you will notice that as the air hits that spot, it will feel cooler than the rest of your forehead. This is a lovely way to set the stage for trance because the middle of the forehead is the place referred to as "the third eye," which is associated with insight and intuition. Sometimes it helps to narrow your focus by imagining the little hairs inside of your nose as "swaying in the breeze" with each inhalation and exhalation.

Good luck!!

Chapter 8

# Honor the Mystery: Invite the Miracle

Self-care is about coping with adversity by seeing you and being you. *Seeing you* means stepping back far enough to gain perspective and discover what you can do to cope more effectively. *Being you* means slowing yourself down enough to feel what is happening within you. Within each of us is the vast potential of an "inner infinity," which reveals itself as the skill of letting go develops. This chapter recounts the ways in which feeling connection to an "outer infinity" can be a part of believing in miracles. When we are faced with adversity, it's natural to look for answers in something bigger than ourselves. Spiritual connection is a variation on letting-go coping.

When you stop and think about it, birthing a baby is a miracle. It has spirituality written all over it. Yet it is not until getting a baby becomes a maze instead of amazing, that we really stop and think about the profundity of how truly miraculous it is.

Until then, you might take conception and birth for granted. You might assume that, like everyone else, when the time is right for you, you will have a child. Not that you are cavalier, but rather when you are in the mainstream, it is natural for practical decisions like where you will live or work, or the scuttlebutt about where to get maternity clothes, baby furniture, and the best stroller to dominate the fantasy, the conversations, and the preparations for the next phase of life.

But, if the mainstream has taken a sharp left turn without you, you cannot help but feel lost and floating like a leaf in a lesser-known tributary. Though floating like a leaf may seem like a pleasant-enough image, under these circumstances, it is anything but; this tributary feels treacherous and ominous—like it is heading for Niagara Falls.

## The Inevitable Question

The magnitude of any life crisis, especially if it seems to be as endless as the fertility journey, is likely, sooner or later, to call to mind questions like Why me? It is universal to look for meaning in suffering. Call it philosophical or call it spiritual, rare is the person struggling with infertility who does not ask this worrisome question.

Worrisome questions are worrisome because they do not have pat answers. In fact, they may not have answers at all. Indigenous cultures understand and respect life's mysteries and pay homage to unknowable forces. Buddhism teaches that control over our lives is an illusion. On the contrary, the larger, more modern, organized religions go a certain distance toward offering solace through answers that emerge from codified beliefs.

This chapter is not about religion per se. When it comes to the place where procreation crashes head-on into spirituality, the need to clarify your awareness of, or investment in, faith can pop up like a jack-in-the-box. The aspect of faith that I am referring to here is more about connection with your inner beliefs and less about embracing or adopting beliefs from religious tenets, although for some, these are congruent.

Growth from the adversity of infertility can pry open curiosity about, or connection to, something larger than us, if it's not there already. For some, faith in God, Jesus, Mary, Allah, Buddha, or any one of the hundreds of Hindu gods can be deeply ingrained and can reflect what has come to feel right. For others, there is a disinterest in, or disconnect from, consciousness about faith of any kind. In these cases, the void can activate, or reactivate, a spiritual search. Those who are clear about their atheism or agnosticism might remain unmoved.

The reality of mind/body unity comes alive in the practice of letting-go coping. When you spend time in the place where mind and body meet, you can't help but notice that nature provides us with a way to find solace. We live *in* but also are *of* nature.

Melissa, an artist, said it to me this way: "Because of my experiences with infertility, I rediscovered nature and spirituality from which I had felt so distant since childhood. It was fascinating to begin seeing myself as part of nature and participating in the great circle of life. I began to realize what an absolute miracle producing a child is. It was always something that I took for granted before. This all made me feel more connected to a higher spirit, the great Creator that is guiding me through this experience, the essence that makes all of this possible." After losing a pregnancy at twenty weeks, Melissa eventually conceived again and gave birth to her son.

## I Have Faith and Hope But Where Is God's Charity?

In forging a working framework for comprehending where you stand spiritually, I subscribe to a concept that Sharon Salzberg lays out in her book

*Faith*. She says, "Beliefs try to make a known out of the unknown. Faith is the ability to move forward even without knowing. Beliefs come from the outside. Faith from the inside."[41] From this vantage point, developing faith is about getting to know yourself in a larger context rather than accepting a larger context that religious beliefs provide for you. I do not believe that these two options are mutually exclusive.

The issue of spirituality is multifaceted. Spirituality can encompass religion. But from a much broader perspective, *spirituality hinges on the power that adversity has to either bring faith to the infertility experience, take faith away from the infertility experience, or both.* Faith infers hope, an important ingredient for anyone who is longing for and striving for a child. Faith and hope are places where the heart must trump the mind. The heart can let go and have faith. The mind is more at home with proof.

Spirituality invites the issue of awareness onto the center of the stage—awareness of something larger than the self as well as self-awareness. Prayer figures into this picture but so does intuition. A common theme among my interviewees has to do with the connection (key word) with the essence (let's call it soul) of who you are and maybe who your ancestors were, with the essence of who your prospective baby is, and with your concept of the divine. The experiences of spirituality described here are rich and varied. The connections that gets called into question are both inner and outer. Wouldn't you say that infinity goes in all directions?

## What Works for You?

A friend once told me that when someone asks him what his religion is, his answer is "trees." And why not? He was really saying that proscribed rituals have no meaning for him, but he honors what he finds awe inspiring. At the same time, keep in mind that every culture has rituals that it considers important. It's just that in times of need, some feel free to make up their own. This is what worked for me, as I explained in the preface to this book.

Jenn captured an aspect of spirituality to which many of you will relate. Jewish by lore rather than law, she said that spirituality to her was "the concept that there is nothing more huge and beautiful and exciting as being someone's mommy." She told me, "I never in my life had a dialogue with God. In fact, I couldn't have cared less. But certainly before my two-year journey led to my surprise pregnancy, I questioned God and asked Him for help. I had spent so much time thinking that I couldn't do something (conceive), which really played with my mind spiritually. It made me feel so insecure and sent me looking for answers in places that I never thought of looking for answers before. I withered a little before I could find the strength to connect with God because I felt so abandoned. In fact it felt like a mind-f*** because I was looking for answers in the same place that I was angry at." Jenn was taking away from the fertility nightmare a connection to faith in a new way.

Cecile's faith came with her into the fertility challenge. If she lost faith in anything, it was in her body. "I felt so betrayed by my body's inability to get pregnant that even when I did, I had to fight the thought that I would not be able to "hatch" this baby." And, in fact, she had a nine-week miscarriage before she conceived again and carried her son to term.

There was a cascade of feelings, from loss of confidence in her body, to loss of confidence in herself, but "not a loss of faith in God." Cecile's main coping mechanisms were to "keep a tight focus" on the mind/body connection. She accomplished this by way of the audio tapes that I made for her, and a ferocious devotion to creating affirmations by cognitively restructuring automatic negative thoughts (ANTS) to counteract her negativity.

Cecile is a religious woman, which enabled her to bring faith to the infertility experience. "I'm Episcopalian. I go to church, and I believe in God." A big part of Christianity for her was having and trusting her faith.

## Lost and Found

For whatever reason, it was easier for Cecile to trust her faith than it was for Nanci, although she and her husband, Joe, were churchgoers, too. For a while, whenever they went to church, she would "just start crying" during the service. She told me, "It wasn't until years later that I realized what the tears were about. It was very painful to be making pronouncements about faith and about believing in somebody taking care of you while simultaneously believing that this will never get better. It was hard to surrender and say, 'Yes, I believe,' when part of me felt like I couldn't surrender to that."

In the final analysis, Nanci felt that her faith had "been tested and she "passed" and [was] stronger now." She went on to say, "There was a dark hour, and then there was light. I relate more deeply now to the church songs and tales, and I understand that it is possible to overcome the concept of the dark night of the soul." For Nanci and her husband, the light came with the adoption of their two sons.

Shari, the event planner turned yoga teacher, is still working on "this higher power thing." She said, "I was raised a Catholic, and now I'm totally into this yoga philosophy. I do think there's a higher power. This yoga training has changed my views on what I thought a higher power was. Now I think the higher power is there, but is within us. If we can tap into that, then everybody's connections will be better." As Shari continued to search for a way to explain to me how she felt, it became obvious that for her, spirituality had to do with becoming the best Shari that she can be.

For Shari there was another aspect to spirituality, a faith that, like Jenn, came from the infertility experience. She and her husband talked all the time about how "uncanny and unbelievable" it was that they got Mai, their adopted daughter. "I don't think there could have been a better match or a better kid if we had birthed her. It's amazing to us. She's funny, she's quirky, she's a goofball, she's smart...she's so loving. She has so many of our traits that we don't understand how this match was made. So yeah, I think it's got to be some higher power or force that made this happen. There had to be because it's just so right. It's unbelievably right."

For Sarra, featured in chapter 5, the faith that went into the experience with her had some ambivalence to it, but faith surely emerged from the infertility experience. On the one hand, she was not sure that God existed. She lived on the basis of "whatever is, is here and now—nothing else." On the other hand, she had a notion that "if a baby is going to come, it's going to come from something bigger than me, and all I have to be is receptive, which I was."

Sarra's was one of the more arduous journeys. She had literally run a marathon a few years before the infertility marathon began. She said, "It's not like there is a route that you will complete in so many hours. As much as I was doing everything I could, I mostly put it in God's hands...because it was just too much of a burden, too never ending for me." Somehow she found the stamina to be a real sleuth for almost four years—to keep on opening the next door, and the next one, keeping hope alive. She threw in a little bargaining while she was at it: "Every time I came back to our apartment, I would kiss the mezuzah and say, 'God, please help me out. If you help me, I'll increase my faith in you.'"

The implications of the fertility quest are profound and vast. So much so, that it is as much a spiritual crisis as it is a biological, psychological, and sociological one. I am not sure that anyone can escape the question Why me?

although some are likely to dismiss the opportunity to use that question for personal growth. The moment you ask, "Why is this terrible thing happening to me?" you automatically slide into trying to answer it, which leads into the option to create a dialogue with the only source who might send you signs that provide an answer—hence, prayer.

## Please, God...

Julie set up an altar that she and her husband prayed at together. She said, "Given the invasive, hi-tech nature of IVF, I felt a strong need to balance that with spirituality." After she opened up to her mother, she most appreciated her mom's friend's attention to prayerfulness on Julie's behalf. And Julie took solace in the fact that "there were a lot of people who were praying for [her] and it made [her] feel like there was a wider circle out there on some subconscious level sending [her] good energy."

Then there was Donna, who was told she had a 5 percent chance of conceiving. She took to "praying hard" for her fondest dream to come true. She now has two children and the second one was conceived naturally. Go figure.

The distinction between prayer and meditation is sometimes fuzzy. Elizabeth Gilbert, in her book *Eat, Pray, Love,* points out one way of distinguishing them. She says that while they both "seek communion with the divine…prayer is the act of talking to God, while meditation is the act of listening."[42] Some would say listening to God. Others would say listening to the god within us.

## Ohm...

I believe that intuition, which has guided many of my patients, can be a by-product of faith, meditation, prayer, hypnosis, self-hypnosis, or the inner stillness that can flow from the relaxation response. Monique, fed up with all of the drugs, began meditating on an "intuitive pop" that she could conceive naturally. After the urea plasma was rediagnosed and effectively treated, she did. Sarra's saga went on as long as it did because she "sensed" that her doctors' dismissal of her as "too old" didn't quite add up. Luckily, she was right.

For some, talking to or listening to God is like an unflexed muscle. In cases like this, when procreation seems so miraculous, and spirituality is little more than a blank space, learning to trust the subtle whispers of intuition can lead to a sense of hearing God's voice and/or trusting that God will hear yours.

Robin's mother was a devout Catholic, but for some reason never imposed her faith on her children. When Robin, an antiques dealer, finally got pregnant, she often found herself thinking about her grandmother and mother, both of whom were deceased. "I felt like they were happy, and I had their blessings with this pregnancy," she told me.

Robin said that she had been uncomfortable with spirituality. Robin's mother died when she was sixteen. Her father had wanted Robin to deny her feelings and did his best to banish her mother's memory. Therefore, suddenly to be thinking about her mother and grandmother's blessings seemed odd.

She found it particularly strange when out of nowhere, in the midst of her infertility, her father told her that doctors told her mother that she would never conceive. Robin was shocked because she is one of six children. Dad told her, "When your mother wanted to conceive, she would pray to St. Ann." Robin told me, "My father doesn't tell me anything about my mother. Why did this information just come to me? So I thought 'Oh, what the hell.' So that was how I started to pray, just imitating the memory I had of my mother praying to St. Ann."

It is curious that Robin's father seemed to "burp" out this tidbit. But it came at a synchronistic time when Robin's openness to doing something new and different allowed her to take on board the idea of prayer, as she said, "in her own way." When life on earth feels painful and burdensome, one place to go is "up and out."

Thanks to the inroads that quantum physics has made into popular culture, the idea of the power of prayer is getting consideration, even by those whose god has been science. There is much from the new age press that is filtering into popular media, but at the time of this printing, a definitive study that skeptics find satisfactory has yet to be published. Believers from the infertility population could care less about scientific "proof." We need to stay tuned on this one.

Meanwhile, spiritual connectedness is enough for believers. Monique described an awesome though "bittersweet" experience having to do with her grandfather, to whom she felt particularly close. She broke down with him and told him about the baby that she wanted more than anything else. Monique told me, "He was eighty-six, and he held my hand and said, 'I love you honey. If there's anything I can do to make this happen, I would do it.' He died the week I got pregnant. I think there's a lot to it." Monique and Bill named their son Walter after her grandfather.

Julie, whose grandma said, "Think a good, come a good," felt that she was able to connect with the spirit of her twin daughters when they were inside of her. "I could also feel the presence of my grandmother when I was pregnant with the girls. I thought a lot about the spiritual side of pregnancy. Ruthie came out first and when they handed her to me, she looked so much like my grandmother it was freaky. She's still like my grandmother. She's fascinated by books, and my grandmother was a huge reader. The girls ask a lot about Granny Graydon. Just the other day Ruthie (four years old at the time) said to me, out of nowhere, 'Granny Graydon died, then I came into your tummy, and then I was born.'"

We live in a secular world. People do not necessarily come out with their spiritual musings—until you ask. When there have been one or more miscarriages spiritual thoughts often pop up.

This leads me to another aspect of spirituality, the working through of grief. Monique had seven miscarriages. Just because the protoplasm vanished, does not mean that there wasn't a connection to the children who might have been. Monique had a way of grieving that worked for her. She dedicated time to expressing her feelings, which she summed up for me as "I loved you guys, and I really wanted you, and I'm sorry. I guess it wasn't meant to be for whatever reason." Monique would be pleased if the souls of one or two of her miscarriages had returned to her in her two sons.

I always ask if there had been a previous abortion. If there is an emotional residue from the past, it is important to know that unfinished business *can* be worked through to enormous relief. The souls of those who were not carried to term seem to give relief when they get respect.

## Your Call If Not Your Calling

Some who are hard-core secularists sound as if they are spiritual in a peripheral way, the way that the moon is peripheral to the earth. Ruth is one such person. She does not consider herself to be a spiritual person. But, she said, "Just watching my son grow makes me think about a million different things...It makes me think about human evolution...and how we've developed into these creatures that are so complicated. I'm in awe that every person on the street was once a baby." For Ruth it is not about God or religion, but what she called a "pagan appreciation of human kindness and motherhood." As an artist, she is "appreciative of what we create and what our minds are capable of thinking about."

Carol and Tom are both declared agnostics. Although intensely grateful for their son and daughter, Carol denies that this has brought her to a connection to a higher power. She claims to be "doing fine" without spiritual ruminations. A left-brain, scientific way of thinking is what resonates for her. Suddenly Carol got quiet as we were having this discussion. She realized how she could express her frame of reference accurately to me. Finally, she said , "If there is a higher power in this, it is Dr. Grifo's brain." Dr. Jamie Grifo was Carol's reproductive endocrinologist and director of the NYU Fertility Center in New York City who orchestrated her two successful IVFs.

While Carol's feelings are as worthy of respect as anyone who might be a religious devotee, I cannot help but wonder if things would have been different for her if she had not conceived on the first attempt at IVF each time. For those like Carol, whose lives are solidly secular, I understand that it would feel like a quantum leap to form a relationship with someone with whom you cannot shake hands. To seek help from this "stranger" in order to find meaning in this nightmare would feel understandably weird. Struggling with infertility does not come with a spiritual prerequisite. Rare is the person who does not find himself or herself giving at least "post" requisite consideration to the topic.

By turning attention to "outer infinity," some find it easier not to give up before the miracle.

## Exercise for Honoring the Mystery

Tune in to the way in which you feel a part of nature. Take yourself outside—no matter the season or the weather. Take a leisurely walk and breathe in the fresh air in your own backyard or neighborhood. Notice the sensation of your lungs filling and releasing. Mindfully, and in slow motion, pay attention to little things that failed to grab your attention until now, perhaps the shape of a leaf, the details of a blade of grass or flower petal. Perhaps you prefer to drive yourself someplace where you can sit at the edge of a pond or riverbank. What do you notice? Lie down in a meadow. Climb a tree. Listen to the birds. Experience yourself as part of nature and its beauty. Be in the present moment with each breath. Allow yourself to feel tranquil, even if it means flipping a negative thought into an affirmation. And if it suits you, pray.

Chapter 9

# Know Thyself: It's All About Awareness

Brenda, a writer and stay-at-home mom said it this way: "The infertility really heightened my personality problems." I would prefer to say that since the unmet longing for a child upsets the psychological balance of who we are, personality traits become exaggerated as we scramble to cope with the uncertainty of infertility. At the same time that infertility disrupts the status quo, there is an opportunity to get a grip on how to be a better you.

## Getting to Know Me, Getting to Know All About Me

Emotions are our responses to our mind/body experiences. Some emotions, such as cheerful, satisfied, and loving are pleasant and easy to experience. Others, like disappointed, pessimistic, or ashamed are unpleasant and more difficult to endure. The full range of feelings is what makes us human. Emotional responses are inevitable, and if we tune into them, they are the place from which we get information about what is going on. We can then choose to use that data not only to find solutions but also to grow and mature.

You would be at a distinct disadvantage if you never learned to notice and articulate your feelings. Evelyn told me that nobody talked about how they felt in her family. In fact, she claimed that she was "actually taught *not* to have feelings." Were it not for the mind/body group and an earlier support group for an eating disorder, she said, "I would never have realized what was 'off' about my family…and I would have continued to squash my feelings." Denial of feelings does not make them go away. Denial of feelings is what can turn them into problems. Those of us who learned that feeling our feelings is taboo had to invest enormous energy in suppressing our realness. How can we stay true to who we are if we are barred from seeing us and being us?

Like Evelyn, Sarra told me that her background had not fostered expression of her feelings either. But at various times in her life, she had the foresight to seek individual therapy and then the mind/body group for the fertility struggle, she said, "so that [she] wouldn't lock things inside." Reaching for help has allowed her to "switch to the land of speaking and expressing and dealing (with issues) as opposed to going into [her] cocoon and covering [her] self up."

To look at your behavior and to feel your discomfort at this time is to put a magnifying glass over your psyche, enabling you to see who you are in the face of adversity. Perhaps it would help to ask yourself these questions:

- In what way am I most emotionally unnerved because of this challenge?
- Am I inclined to do anything about the emotional discomfort?
- Do I find myself exploding or imploding?
- Am I feeling worse both inside my skin and in my relationships than before the infertility diagnosis?
- Am I as angry as the day is long?
- Have I found myself to be marinating in negativity lately?
- Have I become aware that my impatience or insecurities are bigger than the Empire State Building?
- Does it feel as if my fears are waving in the breeze on top of a flagpole?
- Has a chasm appeared between love and true intimacy? If so, why?

Considering these questions can build momentum toward awareness. Awareness will launch the process of change.

While it is true that circumstances have thrown a pie in your face, much depends on how you handle things. Whatever your particular brand of emotional reactivity is in the face of this challenge, consider that it represents you, raised exponentially. Welcome to the club. Given the magnitude of the fertility experience, who would not be in a foul mood? And since there can be a protracted period of time before things come to a logical conclusion, even those who have handled adversity very well usually falter emotionally from time to time under these circumstances.

The plusses and minuses of our personalities fire off automatically. Nothing can change how we experience ourselves and present ourselves without building self-awareness. All of us are expertly aware of how others behave. Noticing ourselves is another story. If we can see what we do under these adverse circumstances, we then can choose to learn how an undesirable behavior can be replaced with a better one.

To most of us, the thought of curtailing a behavior that comes automatically, feels like trying to stop a runaway locomotive. Suppose you can access the mental muscle to stop what swamps you. Then what? What feelings and behaviors do you put in their place, and how? It is fair to say that it would feel like you are reinventing yourself. Then let's say that you succeed in responding differently. New behaviors do not have the weight of history behind them—old feelings and behaviors do. So how does one keep new, more functional responses alive, especially when what is new feels false at first? The complications seem to have complications.

Remember, though, we are hard wired for change as well as homeostasis. What you need is the motivation and the determination to improve yourself at a time when all motivation and determination is absorbed with trying to get pregnant. Eek! Yet, as difficult as this may seem, it is entirely possible.

If procreation had never been an issue, it is likely that another rough spot would have disrupted the status quo eventually. Planet Earth is a challenging place to live. That said, you are in this crisis now, and you have this opportunity now.

It is not easy for any of us to be objective about our subjective experience. Yet, here is where the danger inherent in any crisis can be an opportunity to evolve toward the next plateau of our potential. The idea of evolving toward a higher level of potential can be exciting, and a motivator. It is for you to decide if a new, improved version of yourself is a booby prize or a genuine gift in the quest to enlarge your family.

## Changing the Imprint

Based on our inborn nature, if our early environment was ideally suited to us, who would we become? To a greater or lesser degree, we evolve to fit into our family of origin. The family's way of doing things leaves its imprint. Their expectations control our evolution. The problem is that their sense of who *they* think we are or who *they* need us to be may or may not be accurate.

Imprinting serves a purpose. In animals like ducks, for instance, the first thing that a fledgling duck sees is usually the mama duck. The sight of her sets off a chemical reaction in the brain that mandates that the duckling follow her. This explains why you see duck families walking in line like well-behaved schoolchildren following the teacher.

Imprinting has served a purpose in humans as well. Early man needed the togetherness of the clan in order to survive. The current problem with imprinting is that because of it, we become at risk for feeling trapped in the automatic reactions of an ego, imprinted in our youth, but a reflection of who we were *trained to be.* One of my patients called himself in this process "the domesticated me." Aren't domesticated animals trained away from their true nature?

As our egos form, they become replete with defenses. Egos are demanding little devils. As they emerge, they presume that it is their job to make sure that Tuesday is the same as Monday and Wednesday is the same as Tuesday. They do not want us to get to know our true nature (if, in fact, we've been domesticated away from our true nature) because then the ego would be at risk of becoming "unemployed." Anyone's ego is self-protective, therefore, defensive. It can afford to feel confident, even if it causes us to behave like an idiot because it can count on the fact that sameness has the weight of history behind it. This is why so often we fail to make changes despite our New Year's resolutions.

We become used to our egos, so what they decide feels real and true. But *there is a part of us that can think about what we think about.* That part can evaluate and decide what is not working and what could benefit from change. As the quest for a baby challenges the functionality of our predictable behaviors, the opportunity for realizing that we are capable of change can creep into our conscious awareness (there's that word again).

It is much harder to implement change than to be on autopilot. A more authentic part of each of us is longing to claim what would be a more authentic emotional resonance than autopilot. We have the free will to decide if we want to continue to let our egos hold fast with the arrogance typical of anyone's ego. Our egos will swear that it knows what is best for us even though the ego's old patterns could be making matters worse. Usually, it makes things worse by insisting that what is going on would be better if only someone else did something different.

Our authentic selves know that we must take responsibility for what is happening despite the inclination to blame someone else. Authenticity can get diverted early on if we are misunderstood or insufficiently understood as youngsters. Then we spend a lifetime trying to set the record straight. Usually we attract a cast of characters who, perhaps like your early caregivers, were not capable of understanding you for whom you really were or are. Now you have a chance to reconnect with the person you were *meant to be—the person you would have become in an optimum environment.* This is not to say that your parents were evil. It just means that they did the best that they could, but in some way failed to realize who you were or what you needed.

There is difficulty in allowing a less-practiced part of us explore another way, even if that way would be more authentic. To the ego, this feels like weakness. Quite the contrary, only the sturdy can let go of familiar patterns, stand up to the ego's resistance, and go into the uncertainty of no man's land on the way to developing more functional attitudes and responses. It is not easy to stay steady in the face of uncertainty.

## A Little Guided Tour of the Process of Therapeutic Change

We can best implement change from a place of connection between mind and body. Connection between mind and body is the healing place. The best access to this healing place happens in slow motion with focus on the present moment. If you are like most people, you spend most of your time ruminating about the past or worrying about the future.

Existence in the present moment goes hand in hand with breathing easily, deeply, and regularly. Most of us breathe shallowly, sometimes even when we're not under duress. Stress creates circumstances in which deep, regular breathing flies out the window altogether. Yet, if you decide to "breathe in the now," you can delight in how easy it is to come in for a landing in the present moment. Breathing in this rich way with your eyes closed to screen out external distractions can slow down the way we let our lives whiz past us. Slow motion allows you your best shot at seeing yourself.

Sometimes the cognitive (conscious) and the psychological (unconscious) forces can be at loggerheads with each other. You might *know* that by doing stress-reduction techniques, you can reverse the physiology of stress. But you may not *realize* that part of you may have been *imprinted* to believe that you should not feel relaxed or joyful. This part of you may be out of your conscious awareness, but it will make its presence known by avoidance of relaxation or joy.

For instance, if you've been conditioned believe that you must always have something to complain about, or if you've been conditioned to remain vigilant at all costs, then while part of you may want to engage in stress reduction activities, the dominant ego part of you will pull in the other direction and win the tug of war. You will hear yourself saying something like, "I just can't sit still" or "I'm happiest when I'm really busy." The ego is brilliant at supplying us with excuses.

Resistance to change is virtually universal. Years ago, one of my patients put it this way, "I needed to be dragged, kicking and screaming, into my own happiness." When I am supporting someone in his or her change process, I want to simplify the process as much as possible. I also want to frame things so that the person feels empowered to achieve their goals.

I will often encourage them to come up with a different name for the part of themselves that wants to maintain the status quo. It makes it easier to "see you" if you understand that two parts of you are pulling in different directions.

People catch on right away. The names that they choose have most often been names like bitch, baby, loser, or others names that label the force within them that is stopping their progress toward where they want their lives to go. Sometimes they will give this aspect of the ego a name of someone whom they felt was responsible for modeling the behavior of the troublesome mental imprint. Sometimes they pick a name that offends them in some way—a name like Bertha. What is important is that these types of names have associations that people want to distance themselves from, which is exactly the point.

This is not about blaming anyone. Nor is it about denying that the offending part and its behaviors have had a purpose—they have. We become what we become because we have to. An offending ego part comes into being because it thinks that it is protecting us from some form of disrespect.

Choosing a name for the part of you that behaves in ways that you now want to modify enables you to take responsibility for the forces within, and at the same time clears a space for the purest and best part of you to flourish.

The concept of "ego parts" is useful. In psychological jargon, the term is ego state therapy. It is useful because it forces you to step back and see the totality of what is going on, not just the narrow view of one ego part.

When I use this approach in sessions, it allows me to align with the part of the person who is forward looking and courageous. I am then in position to explain that the battle is now "two against one"—the "authentic" self and me against the person's troublesome ego part. This approach brings relief. Clients can borrow ego strength from my support. Ultimately, my job is to get out of the way; but in the meantime, there's an alternative to the temptation to slide back. By using their alliance with me, a person can shore up his/her courage and get past their vulnerability. It gives them the wherewithal to persist in the face of the inertia caused by the resistant ego.

## Try this Experiment ...

Take a deep breath, which can separate "then" from right now, the present moment. Good. Do you feel things slowing down already, if only a little bit. If not, try another deep, generous breath. Now ask yourself, "What aspect of my personality is most distressing to me at this moment?" Give it a name. Take another deep breath and use the expansion of your lungs to imagine an expanded space into which a pure part of you can inflate. This will provide you with a panoramic view: The troublesome ego part has a separate name, and you have your real name. From this vantage point, read about Anne.

## Who Would Anne Be If She Weren't Always Worried?

A while back, I worked with a woman I am calling Anne, who benefited from separating and naming the ego part that interfered with her goal to learn to relax. Anne was a nervous wreck, yet she felt barred from entering a relaxed state. She exemplified the way an ego part can hold fast, resisting "unemployment." She had a questionable sense of entitlement to putting her true self in the driver's seat. To aim for the relief of achieving relaxation was destined to wreak havoc on the part of her ego that had both primacy and a commitment to anxiety. A "fear-of-unemployment" alarm went off for the ego, and unless I could help her to understand the normalcy of the response to change as "two steps forward, one step back," she would have fallen prey to losing the momentum toward change that she was starting to achieve.

Anne's extreme nervousness was connected behavior modeled by a highly controlling mother, whose mission it was to see to it that Anne never removed Mom from the center of Anne's radar screen. Anne loved her mother and wanted to be close with her, but little by little, she came to understand that her mother was demanding enmeshment. Enmeshment relieved the mother of *her* own anxiety to a degree but hampered Anne from feeling that Anne's life belonged to Anne. Eventually, she became unwilling to donate her soul to her mother. She realized that her anxiety (the symptom) was her inner wisdom telling her she needed to set limits. She named the part of her that resisted taking up yoga as an adjunct to our work "Gloria." We agreed that *doing* yoga would be an easier way to get past the resistance to relaxation than any meditative technique would be.

Anne eventually made it past "Gloria's" resistance to take a yoga class and gradually understood the capacity of yoga to use the body as a doorway to calming both the body and the mind. One class clarified for her what I was trying to teach her about the phenomenon of "two steps forward, one step back."

She told me what happened at the end of her fifth yoga class, when she had come to feel pleased about her new experience of relaxation and her commitment to it. At the end of this class, when the students were instructed to lie on their backs, eyes closed, and just take in the accomplishments of the practice, she said that she "could not get over the depth of relaxation she was feeling"—a true and gigantic victory (two steps forward). She told me that "suddenly, out of nowhere, one of her legs suddenly had a kind of convulsion and started kicking wildly." It was as if "Gloria," the "mother in her," was protesting the change toward serenity (one step back). She understood at that moment that the "Gloria" in her feared that the separation from her anxiety would mean separation from the mother. Two steps forward put her in an unfamiliar place; one step back was the ego's attempt to reestablish equilibrium.

## Separation Anxiety Is Not Just for Children

Anne's story is instructive. Yoga is one way to experience involvement in the present moment and a variation of slow motion meditation by focusing on body postures. But, unless you can separate yourself out from the family soup, seeing and being in the world in slow motion will have its limits.

Stacie and Brian's story is illustrative of this. Stacie was aware that there was a need to set boundaries with Brian's parents. Before a one-week trip to Canada to visit his parents, Stacie told Brian that he "needed to lay the ground rules for what they would talk about." She told me that she was not sure if Brian "would be more protective of them, (as had been customary) or of her."

To Stacie's delight, through the clear communication of her needs, Brian arrived at a much greater level of consciousness about what had been a blind spot for him. I asked Stacie how she thought this had happened. She said that Brian had thought that a pregnancy "would just occur," and all he needed to do was "provide the goods." When a pregnancy did not occur, Brian began to go to appointments with Stacie. He gave her the mandatory injections. The big difference came when Stacie had a frightening experience with hyperstimulation,[43] which "took a real toll on Brian." He told Stacie, "You know it's really hard for me because what happens to you happens to me. When you feel hurt, I feel hurt. And it's hard because nobody asks how I'm doing—not even my parents."

His parent's unawareness of his need for sensitivity may very well have heightened Brian's awareness of his need to be more sensitive to Stacie. Whatever the case, Brian was able to separate from his family's vortex and, in the process, he did not allow his parents to "divide and conquer."

Usually, such a transformation does not happen without a fair amount of therapeutic intervention because of the entrenched nature of family dynamics. Stacie and Brian exemplify how a shared agony can raise awareness, land you in the reality of the present moment, and result in facilitating the change-for-the-better process. Stacie had brought the serenity that she learned in her sessions with me home to Brian. Their capacity for communication developed beautifully. Now everyone is breathing easier. Their twin sons were born in June 2008.

## A Little More on the Science of Change

The need to change is not about what is wrong with you. It is about how your early environment formed your attitudes in combination with how the circumstances of your life influenced you. It is not, by my estimation anyway, about you having a "diagnosis" as anxiety-ridden, depressed, obsessed, or any one of the hundreds of labels out of a psychology text. It is about the behaviors you learned—behaviors that would not have developed unless they had value in your environment. Like all of us, you did your best. With awareness, what you need under *these* circumstances has a chance to evolve. Science is rapidly clarifying the nature/nurture controversy and the news is good. Genetics, of course, control many things. The genes involved in behavior were thought to be as non-negotiable as blue eyes. Recently studies have proven that brain circuitry is more pliable than scientists ever imagined. There was a belief that brain determined mind. It is now clear that the mind can alter the brain's neuronal networks, a true mind over matter phenomenon.

Brain and mind influence each other like body and mind do. Modern equipment—functional MRI, CAT, and PET scans—has shown the brain to be changed by mental training, changed by the power of thought! *Think a good, come a good is not just an old Irish saying.*

When change becomes necessary, it is important to know that it is possible. For the scientific history of how the brain's capacity for change, called neuroplasticity, has come to be tested and understood so far, I highly recommend Sharon Begley's beautifully written book *Train Your Mind Change Your*

*Brain.*[44] Neuroplasticity, the brain's potential for change, is at the spearhead of neuroscience these days.

For decades, studies have shown that talk therapy or behavioral therapy runs neck and neck with, or even surpasses, treatment with psychopharmacological drugs, as can be the case with Obsessive-Compulsive Disorder (OCD).[45] Now, devices that measure neuronal activity have eradicated doubt about the power to use your mind to change your brain. But, psychotherapy is useless without a "mental muscle" investment in making change. Therapeutic change is not passive magic.

## The Next Lap of the Guided Tour of the Change Process

If we were to chart the process of change, it is not a straight line. It is more of a "two steps forward, one step back pattern." *All that matters is that the vector keeps getting farther from the baseline that had been your "normal."* The process can be quite confusing because you make progress and then you feel a gravitational pull backward. The newness of your budding attitudes and behaviors are vulnerable to the ego, which has been insulted and wants to reestablish primacy. The old ego, which you are trying to update, lies in wait, ready to debunk your progress with distortions and other tricks, like Anne's wildly kicking leg. Understanding the process can make a vast difference.

## Let's Get Specific

Coming to a realization that something must change is the "aha" moment that opens the door to your evolution. Julie and others came to understand that their isolation needed to change. In order to do so, the ego's investment in rigidly maintaining privacy needed to give way.

It is common for the mind/body ruckus to open up to "aha" moments. For instance: "Aha–that guided imagery tape really did leave me feeling better," or "Aha–it's been a relief to decide to work less hours," or "Wow–that Tai Chi class really shifted my mood," or "It's amazing that the relaxation response can make such a big difference in such a short amount of time!"

Just as common as the aha moments, is the vacillation that follows as the ego tries to rob the victory from the forward-looking part of us. It is normal to have trouble keeping a commitment to engaging in stress-reducing techniques. It is normal to have trouble to keeping a commitment to making lifestyle shifts.

## Just a Bit More Science

All thoughts have a biological component to them. Remember from chapter 6—our thoughts are electrical signals. The body converts them into messenger molecules. Messenger molecules travel elsewhere in the body and "tell" organs or muscles in chemical language what it is that you have thought.

What impact might the following thought have: "I'm not doing this fertility quest perfectly. I'm failing because I'm a failure." Where might that thought land in your body? These kind of thoughts could be part of an engrained belief system of which you may not even be consciously aware. How can we gain access to the thoughts that lurk underneath our conscious awareness? I beg you—don't be "should-ing" on yourself if you don't know! This is complicated business, and you may need to seek help from a professional therapist. Take the challenge to develop awareness.

The rest of this chapter will bring you into the world of others who have fought this battle before you. It's most likely that you will find yourself saying "Aha," based on a resonance with these people who have so generously shared their stories with you.

## Dealing with Common Ego Parts That May Have Gained "Tenure"

Negativity, anger, lowered self-esteem, impatience, insecurity, and fear are the most common ego states that infertility patients experience. We all have these qualities to a greater or lesser degree. Has your struggle with fertility intensified any of these?

As you read on, you may be inspired to give a troublesome ego state a name other than your real name, if you haven't done so already. This can begin the process of isolating and concretizing the part of you that you can study. These stories should help you to identify what's going on for you.

## Negativity: Saying No to Yourself

We tend to think what we have always tended to think, and if those thoughts are negative, they tend to reproduce themselves when confronted with infertility leaving you feeling weighed down. The emotional response to infertility would naturally pull the rug out from under the most positive personality. Infertility is fraught with disappointment. Therefore, it may not seem as if you have a choice as to whether you let the ANTs (Automatic Negative Thoughts) spoil your picnic, but you do.

Robin, the antiques dealer mentioned in the chapter on spirituality, was both a private patient and a member of one of my mind/body groups. For her, expecting a negative outcome to everything was her default mode. Remember that Robin's mother had died when she was sixteen. Her dad and the new wife, who appeared on the scene soon thereafter, took the position that her mother was dead and gone and that it served no purpose to focus on her. They expected Robin to get on with things as if her mother had never existed. She was told to "close the door" on her feelings. Robin tried to be strong after her mom died. She told me that "[she] realizes now that [she] cut [her]self off from the grieving that [she] needed to do."

When I realized that the family had instructed her not to be human, I encouraged her to bring her feelings of grief out of mothballs and allow herself to be real—retroactively. I helped her to understand that her default mode of negativity was not her fault.

Robin learned to take responsibility for catching herself in her negative thinking, scrutinizing its distortions, and gradually coming to entertain more reasonable, realistic notions. Three cheers for cognitive restructuring!

The net result of Robin's hard work allowed her to dig up a photo of her mother, frame it, and put it in her bedroom. This was a turning point. Her new response to the old message of "don't feel" was to reject that invective. In the process, she became less negative.

Robin said, "I feel much different now. When negative things would come up in the past, I would say to myself, 'Well, of course—that's my life, my rotten life.'"

When we were working together, she did eight IUIs and six IVFs. Her doctors cancelled two IVFs because she did not produce enough eggs. She had two chemical pregnancies and an ectopic pregnancy, which was resolved pharmacologically. By way of laparoscopy, it was determined that Robin had stage three/four endometriosis and her doctor removed adhesions from the back of her uterus.

Of the almost three years of treatment between the ages of thirty-six and thirty-nine before she became pregnant, Robin had fourteen treatments and one surgery. That is what makes her case so interesting. For all of the time that Robin spent singing negative songs, her determination did not falter. Underneath the lamenting, there had to have been a positive engine, much as under Sarra's depression there was hope. The Popeye effect at work again!

"I never went after something so tenaciously like this," she said. "If things didn't come easily to me, I'd be like, 'Well, I never get what I want.' Instead, I was like, 'I can't give up. This is far too important.'" After the fact, Robin said, "I realize I have changed. I created the outcome. I didn't roll over and just accept my fate as I had in the past."

## Negativity of a Different Variety

Cecile, also mentioned in the chapter on spirituality, is an aspiring writer who chose to give up a career in law. She evolved through the issue of negativity in an inspiring and instructive way. She had found out about my work and had flown in from Paris, where she lived, to visit her family and to have a few sessions with me. I made her some personalized hypnosis tapes. Cecile had lost confidence in her body, and she claimed that the tapes were instrumental in holding on to the possibility that she could succeed, "*because* the tapes brought [her] to a place where [she] could experience the connection between [her] mind and [her] body".

Previously, Cecile had become pregnant with an eight-celled embryo and had miscarried. When she came to me, she was in an IVF cycle and was about to have a transfer of a six-celled embryo. Until we made the tape, she thought if she had miscarried an eight-celled embryo, how could a six-celled one work? She said, "I couldn't relax for a second. I felt so betrayed by the history of my body's inability to get and keep a pregnancy that I feared that my body was going to do something besides what it's supposed to do."

She conceived but negativity followed her like a shadow, and she told me that throughout her pregnancy she "had to keep the mind/body connection going." She had intrusive thoughts, about not being able to carry to term. She thought she would not being able to deliver without a C-section. She was sure she would not be able to breastfeed. She carried to term, delivered vaginally, and was still nursing her five-month old son when we did our interview.

For Cecile, beyond keeping the mind/body connection real to her by using the tapes I had made for her, she came up with an original and creative way to use cognitive restructuring, which made a critical difference. Her way of combating negative thoughts was to challenge the intrusion by converting ANTs into affirmations the way you learned at the end of chapter 5. She explained, "Whenever I felt anxious, whenever I thought something was going wrong with the baby, I would go to my computer, and I would type certain phrases and think them at the same time. In the middle of the night, I would just get up, and I would type and type and type until I felt better inside. When I went into labor, before we left the house, the last thing I did was go to my computer and, real quick, just started typing some affirmations."

By the time she delivered, Cecile printed out about four hundred pages of affirmations from the start of her work with me through the birth of her son. With her fingers typing and her mind saying the positive affirmations, she counterbalanced the negative intrusions with an ultimate mind/body activity. Thinking—and typing—a good, came a good! Thinking and typing a good was the new behavior, which was able to withstand and override the negative thoughts when they intruded.

Cecile is aware that the scars of her infertility and her negative outlook are still there, although she is a lot less negative. It helps to know that she has the power to convert her angst into affirmations. She will continue to work on this so that she can continue to invite the negative ego part out of the driver's seat when it appears. She is clear that her inclination to think negatively is not an indictment about her, but rather a bad mental habit. She knows how to work with herself and even in labor, had the presence of mind to type positive affirmations. For Cecile, negativity is her growing edge. None of us is ever a finished product.

## The Energy of Anger: Expletive Deleted

Anger is a gigantic emotion. Whether we feel it, express it, deny it, or stifle it, it can have a volcanic force. Unlike a volcano, we can harness the energy of anger.

In its pure form, anger's magnitude stands on its own. Anger can also be a disguise for emotions that are more difficult to experience. Anger can be a derivative emotion, a catchall place for other difficult emotions to take refuge. Anger is easier to experience than sadness, loneliness, or grief, for instance. These emotions are painful and it's not uncommon to convert those emotions into anger.

Sometimes we can turn anger into something that we *do.* It's easy to go into action, railing at whomever or whatever, blaming, cursing, or fighting. To a certain extent, this relieves pressure, but only as a safety valve. It does not solve anything.

If you are going to make lemonade out of the infertility lemons, stepping back with mindful sobriety is required. Anger can be blinding. If you can catch yourself in an angry outburst or mood, and be responsible to it, you can step back and ask yourself what would be more productive. This is truly an opportunity to learn how to manage anger differently.

From 2002 until 2007, I gave out a questionnaire before the first and after the last session of my ten-week mind/body stress-reduction groups to forty-nine participants who were willing to devote forty-five minutes to filling it out each time. I was looking for the impact of group participation on physical symptoms, mood, and levels of optimism of these participants. The questionnaire included three self-rating scales,[46] which were designed to measure these aspects. A researcher, Kris Bevilaqua, PhD, interpreted the data for me.

It was determined that at the end of the ten weeks, the category of somatic complaints had trended downward. Life orientation statistics revealed that feelings of optimism had gone up. The mood profile yielded a most happy surprise. Of the moods measured, anger had decreased to a level that was *statistically significant!*

So when I say that to change your emotional reactivity is possible, be inspired. It feels lousy to be angry, especially from the body's perspective. At the same time, anger is a most logical emotion.

The breadth and depth of anger varies, often coming entangled with other emotions like jealousy or in response to insensitivity. For instance, Kelsey, an elementary school teacher, continued to be angry even after she succeeded twice with IVF. She remained resentful and bitter whenever she heard a story about someone who conceived easily. While it was true that her emotions simmered down, she told me, "A lingering bitterness remained toward those who didn't have to work like I did. Despite my awareness and appreciation of how lucky I was, and my genuine happiness...that my friends were having babies I still couldn't shake the fact that things had been so easy for them."

One day, when she was discharging venom in this way, her husband stopped her dead in her tracks when he said, "If I had it to do over, I wouldn't change a thing." At first, she thought that he was crazy. She wanted to "erase all of that pain and suffering." Then her husband said, "If it had happened any other way, we wouldn't have Evan."

She went on to tell me, "Something about that simple statement really shattered my previous way of thinking. He was absolutely right—if it had happened at any other time...we might have a child, but it wouldn't be our daughter, Evan. And our second child wouldn't be Theo. All we had to endure led us to having the perfect children for us. It made me think. That must be how it is for everyone after going through infertility and finally getting a child by whatever method. I think about that conversation a lot, and it makes me realize that if I had the choice, I wouldn't change a thing either." The silence on the other end of the phone was palpable. Then she said, "Wow, I never thought I'd say that."

This example is not so much about coming to an awareness of an emotional response and letting there be a shift in attitude. It was simpler. Her husband provided a reframe. Yet unless Kelsey wanted to let go of the anger and bitterness—wanted to evolve emotionally—she might have chewed on her husband's comment but would have been likely to spit it out.

Sometimes anger is totally pure and uncontaminated by other emotions. That was the case with Jenn, the singer/songwriter. She spoke of the anger she felt when she was preparing to adopt. She said, "I thought of the social worker coming to my house to make sure that we were going to be good parents. I felt that I should be exempt from that. I just went through two years of goddamned infertility treatment, and somebody else down the block smoked through her entire pregnancy. I haven't put one thing in my mouth that I'm not supposed to be putting in my mouth, and you're asking me if I'm gonna be a good parent or not?" Her rat-tat-tat sentences were a function of her anger, exasperation, and sense of injustice.

Acknowledging anger was healthy, and its expression needed to happen in order to get it out of her way. Nature loves the truth and if you are angry, it is healthy to express it. What is critical here is to separate the anger that shows up automatically in response to bad news or disappointments, from the anger that zooms out of what Wally Lamb calls the "museum of justified indignation" in his book *I Know This Much is True*.[47] Ask yourself, Is there an emotion that is stored in your psyche, on which anger may be piggybacking? Is this anger activating an old, unresolved issue?

Anger is a normal emotion, a built-in aspect of anyone's humanity. In your case, it might be an ingrained and troublesome aspect of your personality that infertility provokes and legitimizes. On your path to emotional growth, you are in position to discern if for you, anger could actually be a troublesome, long-standing trait with an excuse to make an appearance. Anger can be a subterranean handicapper of well-being. Now is your chance to rework your neural wiring and, by so doing, escape from the burden and discomfort of living in an angry body.

For Monique, anger was a double-whammy. She experienced the common theme of anger at the betrayal by her body's inability to conceive and anger at her heart due to a congenital heart defect.

Nevertheless, Monique knew that anger was her "default mode." Monique became aware that anger as an ego trait was something learned from her parents, a kind of emotional inheritance that was part of her as if it were genetic. She said, "I thought anger made me a strong person." She had learned "to be a control freak and got mad if things didn't go [her] way."

Anger at her body (for two reasons) and anger as a default mode eventually provoked a triple healing. Meditations had become one of Monique's mind/body tools. She had learned that "you sit with yourself and nothing happens, nothing happens, nothing happens, and then there's a shift."

In one of Monique's meditations, she was inspired to picture herself as a little baby when she was pure and beautiful. Suddenly she started crying. Soon she realized that she had been angry as a baby because she knew intuitively, even then, that something was wrong. She knew it was her heart. This important recognition opened the floodgates and released the locked up emotion, which was *a long-standing fact that her body knew*. The connection of her body's experience to her mind released a tamped-down truth and freed her in a new way.

Cumulatively, meditation facilitated Monique's transformation of the anger issue in general. She said, "When you are screaming at someone, your adrenalin is up. You have a puffed up sense of yourself, as if you are more powerful than you really are and that somehow people should be scared into giving you your way." But, the more Monique meditated, the more "[she] realized that [she] started to feel an inner strength...[She] didn't need to be angry because strength comes from a clear sense of who you are and not from whom you can intimidate."

Now Monique can say how pleased she is to be dealing with the issue of anger. She is relieved to have the tools now. "Dealing with anger," said Monique, "leaves more room to be open to joy, another by-product of getting to know myself in a more profound way." She was pleased that she had learned to short-circuit the flow of adrenalin when she was pregnant and pleased that her shift away from anger as a default mode would "make her a better mother."

## Self-Esteem: Putting Humpty Dumpty Together Again

Humpty Dumpty is an egg—no pun intended. He has a great fall off the wall and breaks. Unlike the nursery rhyme, which champions collaborative nay saying, you *can* put your Humpty together again. All of the king's horses and king's men are irrelevant. Nay saying is the last thing that you need right now, yet the echoes of negativity may seem to be conspiring to convince you that all is lost. With that, your ability to esteem yourself flies out the window.

Negativity and anger can mangle self-esteem. What is required in order to rebuild or rework your self-esteem is self-esteem itself. This may seem both paradoxical and impossible. Yet even if the emotional blow of infertility has made you feel like you are down for the count, if you are breathing, you can come back stronger than ever. Each breath is esteeming. Many breaths are self-esteeming.

Those of us who have had a habit of not esteeming ourselves were likely to have been inadequately appreciated as youngsters. Some of us are left feeling as broken as Humpty Dumpty. We are all vulnerable to large and small insults that alter healthy development.

Put in a context that is based on current science, Giacomo Rizzolatti, in the early 1990s, determined that there are neurons in the brain that have come to be known as "mirror neurons." These neurons are critical in the development of language (communication) and emotional awareness (feelings). They are activated in specie-specific relationships such as when mother's and child's eyes engage, and as the baby is noticed with delight and approval.[48] The baby's ability to have impact on the parents is what builds the esteem to believe he can have impact on his world.

It is not hard to understand how important it would be in a baby's development to be mirrored, which is what happens when the mother reflects recognition, delight, and approval back to the child. Furthermore, our neurochemistry and the exposure of our brains to hormones as we are mirrored are what ensure that we develop the connection to others that we then use our mirror neurons to experience. Our sense of ourselves comes from how others show us what they see. Unfortunately, in our early years, if others show us a distorted view of ourselves, we do not have enough of a perspective and cannot help but take those distortions onboard. Self-images, thereby, become more like reflections in a fun house mirror, reducing, if not greatly reducing, self-esteem.

Despite the imposition of a distorted self-image from the outside, self-esteem is not set in cement. Letting go of the mind/body turbulence of infertility by taking yourself to that inner or outer infinity where you can be you and see you, goes very far toward putting the Humpty in us back together.

## Impatience: I Want My Baby Now!

Brenda called them personality problems. I prefer to reframe them as personality traits or ego states that become more obvious due to the infertility battle. They can mess with your level of self-esteem. On a bad day, the distortions seem like gospel truth. Don't be fooled. Feelings of "I'm not okay" are feelings that we all can easily cook up while under duress. These, to my way of thinking, are areas where your cognition could use some restructuring.

What you want to be over and done with can seem eternal. You wait for test results, wait to heal physically or emotionally before the next attempt, wait after varicocoelectomy to get "the verdict," you wait for the hCG surge for an insemination,[49] wait for news of fertilization in a Petri dish, wait for your period, and—the big one—the two week wait, affectionately known as the TWW, to find out if you're pregnant...good grief! No wonder you have turned up impatient. Virtually everyone does. How could it be any other way?

Remember Lori from chapter 7, who looked back at the two years that it took her to "get to the starting gate?" Remember her impatience when another ovarian cyst and more endometriosis required another surgery? In addition to feeling "antsy, bored, and dissatisfied," she felt the impatience in her body as well, noting, "There was heaviness in my chest, and it felt as if my legs each weighed a ton." Mind you, she weighed ninety-five pounds soaking wet, yet she felt "trapped in her body, finding it difficult to move and difficult to stay still." These were the remnants of what had been depression and sky-high anxiety at first. Lori would be happy to tell you that one of the gifts of the fertility quest for her was learning to "breathe in the now." So much about these circumstances is frustrating. Developing patience is part of developing acceptance of what is. It is also a form of self-respect and self-kindness. No amount of impatience will make the time go faster.

It is bad enough that assaults come as part of the process. How unkind to assault yourself with the mental machinations that flow from impatience. We can learn to hang on to the best of ourselves—one breath at a time. It may seem like a cliché to continue to refer to the breath. It is important not to lose sight of the fact that the breath actually *is* a tranquilizer. Using the breath to achieve even momentary tranquility, in combination with cognitive restructuring, the relaxation response, and self-hypnosis, can bring us to a place of feeling empowerment, as we realize that we can manage our ego states and personality traits.

## Insecurity: What If...

The change to a frenzied state of mind and body because the edict "infertility" has come down on your head opens up to a vast unknown. To feel beset by insecurities is logical because when the vast unknown opens, a no man's land of uncertainties cannot be far behind.

Brenda, who now has three little girls, talks about the evolution of her awareness of her insecurities. When she was going through the challenge to conceive the first time (her two younger daughters resulted from "freebie" surprises), she felt as if she was in a "war zone." When she was in the war zone, she was geared for battle. It's only after she got her family that Brenda came to feel the "whiplash" that has left her feeling "wobbly in her self-esteem."

Now, Brenda says she is feeling "insecure about certain things that [she] never felt insecure about." Her childhood had been tumultuous and damaging, and much to her credit, she followed through on the idea to learn mothering skills from an expert. Even before her first daughter was born, her intuition had led her to be proactive and take classes in childrearing. She has changed all of her inherited rules about parenting, rendering the mother imprint in her ego "unemployed." Payback from her "unemployed" ego has come in the form of insecurity, even as the childrearing classes should have shored up her confidence. Believe it or not, this is logical, because by leaving behind the familiar, even if the familiar is awful, the uncertainty of the new invites insecurity to zoom in. Two steps forward, one step back. Eventually, new behaviors become more solid.

Brenda is suffering with her variety of insecurity *for the right reason;* she is giving her daughters the kind of loving attention that she never had. This is bringing up what she calls "stuff from my childhood that was buried." Brenda is facing and navigating through her "stuff" to get to the other side, rather than giving up or avoiding it.

In another sophisticated move, Brenda and her husband gave themselves geographic distance from her dysfunctional family of origin. Brenda is pleased to have a "safety zone," that is allowing her to feel and deal with her insecurities without the constant bombardment of the chaos-creating cast of characters. Moving away from problems does not solve them. In Brenda's case, however, the safety zone has allowed a panoramic view that is serving her emotional growth very well. Geographic distance has made it easier for Brenda to *be* herself and *see* herself—and grow.

It is most common to feel insecure in the midst of the fertility struggle, where decisions and choices can be so confounding. Brenda's insecurities became more prominent after her infertility was resolved. Like others in this chapter, Brenda has tools that will help her to navigate through life's challenges.

These tools are in this book for the taking. If these tools are not enough, will you be able to get your ego out of the way so that you can reach out for professional guidance? We do not think twice about leaning on crutches while an ankle or knee heals. Emotional conditions can be healed as well, and just like a bone, can be stronger in the broken places.

## Fear: When the Bug-a-Boos of Boo-Boos Go "Boo"

If the uncertainties of infertility do not provoke insecurities, I will eat my hat. If the uncertainties of infertility do not provoke a fear of the unknown, I will eat my hat without salt and pepper. One of my patients put it this way: "I know a lot of people who find all of the tests and anticipation before the diagnosis to be the worst part. There is a whole infertility language that you have to learn and that can induce fear." And that is only for starters.

Fear. What a concept. When applied to infertility, it is not the same as anxiety. However, if you are anxious to begin with, fears can merge synergistically with underlying anxiety and make your life miserable. Fear can also be your teacher.

Negativity, anger, low self-esteem, insecurity, and impatience are the common personality traits that balloon in magnitude for many people. Fear cannot help but be part of the picture. To a greater or lesser degree, everyone's hair stands on end at the prospect of medical procedures. As one woman put it, "All of a sudden you are hearing words you've never heard before, you're wondering what you will find out today, and in anticipation of being poked and prodded you worry 'will it hurt, will this work, will my insurance cover this?'"

A common fear is the fear of the high doses of medications that stimulate ovulation. The jury is out on whether these fears are misguided. The medical consensus seems to be that the risk is low. A majority of women decide to trust the process. After all, the drugs do help the body to function in a way that it has not been able to do on its own. After several IVFs, some begin to question the impact that the drugs could have on health. At this point, some women switch to a gentler, more holistic approach. Some women go for the more holistic approach right off the bat.

The fear-of-drugs is a risk versus gain story. It is only one of many decisions that you need to make. The need to sift through complexities is one of the things that make the infertility experience so emotionally arduous. For this reason alone, self-awareness so that you can communicate with yourself and others is critical.

## Under the Turbulence

Negativity, anger, low self-esteem, impatience, insecurity, and fear responses are the turbulent manifestations of the mind/body experience of infertility. As you learn to monitor how you react, you can decide how you want to respond. With a clear sense of what is going on, you put yourself in position to dive under the turbulence to get relief and to get solutions.

This chapter exposed you to six major personality traits that commonly escalate in the fertility quest. The mind/body exercises, especially the relaxation response, mindful breathing, and self-hypnosis should be revisited and practiced. They are what take you under the turbulence where you can *be* with yourself and find relief.

Chapter 10

# Gain from the Pain: Would You Believe There's an Upside?

So what do you stand to gain by suffering through the delay in getting your baby? Your cheerleaders, who are on the other side of the infertility battle, are eager for you to know what the experience meant to them so you can feel inspired to hang in there. This book has been, among other things, an invitation for you to ponder how this experience could turn out to be surprisingly beneficial. Read these stories. Which resonate?

Growth from adversity always involves coming to terms with something that you would never have chosen. Looking for what you can gain from any adversity is not what concerns you at first and may even annoy you. Yet it is a way to assign meaning to your life and avoid living with bitterness.

## Not Just Strength but Inner Strength

When I was a little girl, my grandmother used to use an expression that I did not understand at the time. She would say, "You should do such and such—it will put hair on your chest." Hair on my chest? I was preoccupied with the not-so-hidden message that she thought it was better to be a man. Eventually, I understood that she was saying more than that. She apparently thought that men had a corner on the strength market. On another level, "such and such will make you strong" was the communication I was supposed to derive from her words. There are other less judgmental expressions of the "what doesn't kill you will make you strong" ilk.

The issue is not just about getting strong. It is about feeling strong, *owning* the strength that can build in the face of challenge. The life force, the Popeye effect, whatever you want to call it, is a hard-wired aspect of our nature. It is within us, but some of us are more purposeful about developing it than others. Developing inner strength can be both a conscious and an unconscious by-product of adversity.

Melissa, an artist, put it this way: "If it had not been for this amazing challenge in my life, I would still be afraid of the great unknown and would wonder if I had the balls—I mean ovaries—to get through it. I now know that I can and will get through anything."

Unfortunately, some of us are born into environments where developing inner strength is not encouraged and may even be discouraged. This kind of environment can rob us of the drive to feel and use our capacities, leaving us likely to form an inaccurate picture of ourselves. Personalities, or aspects of our personalities, form around distortions. When adversity brings us face-to-face with ourselves, we have a chance to correct our course. All of us get tossed around by life. As Gilda Radner once said, "If it ain't one thing, it's another." My point is that with awareness, we can set the record straight if our sense of ourselves became distorted.

Self-awareness can open us up to *what needs to be changed*. As you continue to navigate turbulent waters, self-awareness can bring you to a realization of *what has changed* because of your efforts. Reveling in the self-awareness that develops cannot help but call attention to increasing levels of inner strength. In the process, we stand to discover or rediscover who we were *really* born to be and consequently connect with our in-born authenticity. Inner awareness and inner strength make for a wonderful partnership and form the substrata upon which gains from pain accrue.

## The Heart of the Matter

Seeking authenticity or connection to our in-born realness does not mean that we have been inauthentic. It just means that the lessons that come from the impact of unavoidable stress give us a chance to evaluate what feels right and what does not. It is up to us to recognize and honor the messages that bubble up from the inside. Can you be honest about aspects of your lifestyle that are not working? Can you see your stress warning signals as gifts? Recognizing these messages can be challenging. They can be quite subtle. Sometimes we don't have access to our true selves. Sometimes our suffering can block access to hearing that inner whisper. Sometimes we don't hear what is coming from within, even if it screams at us. As Winston Churchill once said, "Men stumble over the truth, but most of them pick themselves up and hurry off as if nothing ever happened."

Realness is simple when we are infants. When we are hungry or uncomfortable, we scream. When we are afraid, we scream. When we are content, we are free to vocalize and play with abandon.

As we get older, with years of experiences stamped on our templates, we gloss over our inner knowing and freedom to express how we feel. The infertility diagnosis all but guarantees that you will get bumped off track, even if you are usually in touch with what you are feeling, Now you have a chance to quiet yourselves and learn to hear or see or feel—and trust—the whispers or shouts from within that can put you back on track. *You will feel the resonance of your truth if who you are is congruent with where you are going.* The synopsis of how others gained from their pain can be a beacon, shining on what you can gain as well. Read on.

## Ellen's Gains

Ellen, a photo editor, called me when I had already written seven chapters of this book. "Was it too late to participate?" she asked. I gladly set up an appointment to speak with her.

When I opened the door, I noticed immediately how well she looked. Her facial features were soft and relaxed. Her twin son and daughter were fourteen months old and she was back to her very challenging job. Yet she looked younger than her forty-two years and younger than she had looked when she was in the midst of the infertility crisis.

Ellen told me that she had a breakthrough moment recently that made her say to herself, "Oh my God, I want to contact Helen and be a part of her book. All of a sudden, I realized that I am using all of these things that I learned. I've grown from this experience.

I realized the incredible joy that has resulted from our pursuit of this goal. It is a miracle. Miracles are possible if you really set your sights on them. I am joyous every minute that I'm with the babies and never forget that feeling when I am away from them." No wonder she looked so good.

This breakthrough came at a point in time when Ellen had been feeling stressed and tired from her two full time jobs—work and motherhood. Yet she called me when she realized that meeting the demands of this combination of stressors was possible because the self-awareness tools that she learned were serving her well. Ellen called me because she wanted to share her excitement with you, the reader.

In her twenties and again in her thirties, Ellen had participated in Outward Bound. They had been the biggest challenges of her life. Now she understood that infertility "was like Outward Bound, in that it strips you to be face-to-face with yourself and shows you your inner strength." She determined "that infertility was the biggest Outward Bound of all." I might add that it can also be the biggest inward bound experience if you let it.

Ellen also wanted you to know that "when you are at the beginning of any challenge, it is never obvious which path you should take." She began her quest to parenthood at thirty-nine. Herbs and acupuncture did not bring her FSH down. Clomid,[49] and inseminations, got her nowhere. Ultimately, the third reproductive endocrinologist and the second ovum donation cycle was when she hit the jackpot. Her babies were born when she was forty-one.

Authenticity for Ellen cuts a wide swath. It resides in the awareness of her inner strength, in an unshakeable resolve to do everything possible to get to any goal, and in never letting herself move very far away from experiencing joy. Along with joy has come an intense love. This struggle really opened her heart to the babies and her husband in ways that had been unimaginable.

Ellen also takes great pleasure in the awareness that her level of self-esteem has risen. She has achieved a belief in herself *and* a faith that if she needs help, she can get help. If she has one regret, it was that she did not reach out to me for emotional help sooner, now thinking that the struggle might have been shorter.

An important aspect of living from a place of authenticity for Ellen that she wanted to be sure I shared with you, was the importance of acceptance. "I realized along the way," she told me, "that people who are successful don't keep trying to do something the same way when it doesn't work. I had to step back from myself and look at the bigger picture with flexibility. I accepted ovum donation, and I was prepared to accept adoption if need be."

## Jenn's Gains

Jenn, the singer/songwriter, is a hard-core advocate of her belief that realness cannot be faked. She described the clarity she gained about this while struggling for a conception. She did a great job of keeping herself positive during her early IVF attempts. She said, "I felt as if I were a boxer. I went into the ring and I went through one round, and I was still standing. I had some cuts and bruises and my ribs were hurting me a little, but I felt I could get through this. But by the third round of IVF, I was starting to wobble. I didn't want to care, but I could not fake nonchalance. The fourth IVF knocked me out. I had to get down on the floor and cry 'uncle.'"

Jenn could not fake readiness for adoption before she actually was ready. She collected all of the information to pursue this path but realized that something within her was saying no. The knockout punch of the fourth IVF failure allowed her to let go of this path. When she emerged from her depression, she said to her mother, "Mom, I'm ready to adopt. Start knitting."

What a shock when pursuing adoption simultaneous with using a surrogate for the remaining frozen embryos came to a screeching halt with the unlikely and natural conception of her son. This conception was in defiance of her multiple diagnoses and state-of-the-art treatments that anyone would have considered science fiction as recently as a decade ago.

Before the birth of her son, Jenn remembers feeling incredibly grateful for being "sick as a dog" for the first four and a half months of her surprise pregnancy. She said, "I could hardly talk on the phone without feeling that I was about to throw up. I'd go into the bathroom, and I'd be dry heaving and I'd giggle and say to myself 'if I didn't have to work so hard for this I would be really resentful right now.'" I can say that gratitude was at the top of everyone's list of gains from the pain.

In the midst of her depression, Jenn cut off her beautiful curly red hair for Locks for Love. Contemplating surrogacy, she thought of "the woman who would be willing to be nauseous or perhaps get gestational diabetes so that I could have my child." Although in no mood for short hair, it seemed like a small sacrifice compared to the much larger one someone who did not even know her would make for her.

For Jenn, altruism looms even larger now. Concern about the environment has been important to her. She has written a number of songs for a not-for-profit children's group. "We're now trying to get celebrities and corporations to get on board to help get the CD into schools and other places where kids can start learning about the environment in a nonpedantic, nonscary way."

Jenn was relieved to replace two years of feeling sorry for herself and talking about "her body and her hormones," with something that serves the greater good. For her, this would serve the double purpose of showing her son that he can be proud of her work ethic and emulate her example.

Before Jenn was in her own quest, she could not understand why her friends did not move to adoption as quickly as she thought they should. Now she knows that you cannot fake getting to a decision until you get to that decision. Another gain for Jenn was the realization that she could be quite judgmental. "I've learned to be more forgiving of how other people walk down their path. And I've also learned to be more forgiving of how I walk down my own."

## Monique's Gains

Monique can count ten—ten!—benefits from her infertility ordeal. The first, which led to her pregnancy, was when she began to listen to that inner whisper that eventually screamed, "You do not need fertility drugs!" This coincided with her doctor's recognition that she and Bill needed to be on antibiotics simultaneously to deal with the infection that they were passing back and forth with intercourse.

The second benefit was the incredible friends who came into her life, a benefit so many others would hardily agree with.

Third was the shift to be an infertility mentor. She is planning a book about infertility so that women understand that "not all cramps mean that you are having a miscarriage." And, lucky for me, she was the first to sign up for my project.

The fourth benefit was listening to the whispers and shouts that emanated from her regarding her profession. She allowed her unhappiness in her job to grow in awareness. She had abandoned her law degree, and then she abandoned a job related to her MBA. She discovered that her passion lay as a volunteer teaching entrepreneurship to inner city high school kids. "This got me out of my little world to a place where I could revel in the cool way these kids respond." She had listened to her unhappiness. As a result, she gained congruency between her heart and her occupation.

She has given herself permission to follow her passion. To achieve this, Monique needed to free herself of what she called Catholic guilt. Her experience had been "if you were happy, something was going to get you." She now sees this as "bunk." This is her fifth victory.

Sixth, she became aware that all of this inner listening could lead to the transformation of anger. She shifted away from being a control freak. She forged a path inward to serenity that she travels on a daily basis for renewal. Having mind/body tools for life was a universal theme on many interviewees' list of benefits.

Seventh, she opened to joyousness, which she was clear had come as a result of living in the present moment. Realizing that if you are okay right this minute, or if you can get to be okay right this minute with a letting-go technique, then that is as good as it gets. The future is a question mark for all of us and the past is over.

Eighth, she is clear about the growth of her empathy for others based on understanding the challenge to get pregnant and the worries about being pregnant. So, dear reader, people who have been there, have compassion for what you are feeling.

Ninth, she learned the Harville Hendrix style of communication technique in the mind/body group, which she and her husband continue to use to their great satisfaction. (You learned it at the end of chapter 3.) Their relationship has deepened. Communication is the key to deepening intimacy.

And, tenth, she was able to cut negative people out of her life. She is pleased to have the discernment and assertiveness to screen out people who are "so self-involved that they only have a rigid awareness of their own struggles and lack sensitivity to mine." She laughs that she needs to make an exception for her mother-in-law.

For Monique, the essence of all of her gains is that adversity is not just something to get through, but also something to grow from on a lifelong basis. She is very grateful for these benefits.

## Brenda's Gains

"I speak my mind more," Brenda told me. It is one thing to learn how to see who is supporting you in a good way. It is another to learn that you need to be a better support to yourself. That meant that Brenda would have to keep certain people at arms distance. "I never got what I wanted from my family, but I'm finding that I'm getting what I want from the new people in my life." She has learned to assert herself and take into consideration that some people are unable to respect her needs, no matter how clearly she communicates them because they are totally preoccupied with themselves. Sadly, some of these people have no interest or capacity for growth.

Furthermore, it is a thrill for her that, because of her hard work, she no longer feels like a victim. "I feel much more confident in my own opinion, even if it is different from everyone else's. I was never able to assert myself before." This is a bigger deal than might be thought. Victimhood creates a centripetal force, swirling in on itself and reproducing itself as if on automatic pilot. Brenda's escape from victimhood is a huge victory.

Her awareness that "she didn't want to repeat history" made it possible for her to enlist the help of those who could guide her to her goal of being a great parent. She started taking childrearing classes as soon as her daughter was born. This has brought her enormous satisfaction and relief.

Brenda had three successful pregnancies that were fraught with terrifying circumstances. She had low amniotic fluid, which limited her freedom to be active. With one of the babies, there was a scare about the possibility of deafness. "Every time I went to the doctor, they told me bad news, like 'the baby is not growing properly.'" She herself had a blood disorder and insulin resistance. Through three pregnancies she was on the edge of her chair and in the end, the outcome was excellent. Mind/body techniques, which she now has for life, helped Brenda deal with the anxiety.

## Lauren's Gain

Lauren's reconnection with her gut instincts were similar to Monique's in that she felt that she could get pregnant on her own and carry to term. Her "infertility" was of undiagnosed origin, and she said, "It was like cognitive dissonance to be told that I had a five percent chance of ever getting pregnant, and I should just give up and put my money into adoption, even as I was conceiving."

Lauren and Kevin did IVF five times, conceived once with twins who miscarried and once with a biochemical pregnancy. When she finally decided to trust her gut, she abandoned IVF and conceived on her own four times, and miscarried three times. The fourth time resulted in her son.

Staying connected to her gut instincts was not easy in the face of medical nay saying. Once, during an embryo transfer, her doctor told her that the embryos were not looking particularly good. "The doctor actually said to me, 'Well, let's see what you can do with that little eight-celled thing, as she waved her hand in the air and walked out.'" Thing! This story almost defies belief, but it is true. As an aside, it raises the issue of the wonders of modern science when it is separated from the wonders of the human heart. But back to Lauren's gut instincts.

Keeping her authentic connection to herself had another aspect that Lauren felt could have been even more important than believing she could conceive and carry to term without medical intervention. Lauren's mom was consuming and controlling. Mom had instilled in Lauren a feeling that no one would love her as much she did. That Lauren could fall in love with and experience love from Kevin, a delightful and appropriate husband was a small miracle.

She had been convinced to feel obligated and loyal to her mother to the exclusion of forming meaningful friendships. "My mother always made me feel bad about myself. I was never doing enough. I was never pleasing her enough. I was never working hard enough to do whatever it was that she wanted." Lauren was an only child, an adoptee whose inappropriate "job" it became to satisfy the insatiable woman who happened to be her mother.

Lauren opened up to beautiful friendships in my mind/body group. "I realized when I joined the group that I had not had the energy to work on other relationships given my mother's demands. I realized what a relief it was to have these wonderful women friends now."

There was a pivotal moment when she came in for an individual session. Her gut had told her that her mother was somehow inhibiting her from having a successful pregnancy. She ruminated over this with her friends, but it was not until she heard me say in that session, "It seems as if it would be hard for you to have a baby, when you already have an all-consuming child on your hands." When I said that, I watched her melt with relief. This had been her hunch all along. Soon thereafter, she conceived and carried her son to term.

Lauren was able to turn her relationship completely around with her mother. She was able to transition from the brainwashed position of "maybe I'm being too hard on Mom" to "Mom has lost the power to push my buttons. Now I can see what she is doing and say, 'yeah, whatever' to myself. And I realize that all along there was an underlying anger toward my mother, which is not there anymore since I established firm boundaries."

Predictably, her mother has found other people on whom to push her demands. Lauren came to trust that she could not only conceive on her own, but also deliver a healthy baby. Furthermore, she gained the freedom to be appropriately available to her son.

## Carol's Gains

When I asked Carol to assess the gains, she practically shouted with joy, "My marriage!" She told me that their "honeymoon came to a screeching halt, but ultimately it brought us to a more mature, confident, empathic, harmonious, and strong place in our relationship. Nothing can polarize us now." For Carol and Tom, rejecting the need to "feed starving people in Africa," their hearts opened to each other in new and beautiful ways. Like so many others, they got beyond the barrier to communication based on her need to talk and his aversion to it, and as a result, communication became the path to rewarding intimacy. How exquisite to be inside the experience of intimacy.

## Cynthia's Gains

Cynthia, a dance instructor who had been a member of one of my mind/body groups, waxed philosophical in a long e-mail to me expressing the ways in which she "changed her perspective" about being who she was and "the many lessons" which she knew she would carry "forever."

She wrote, "I never knew how strong I was until I gained power over my body and mind. The group experience was huge for me. We were all in the same boat, supporting each other, and I learned how to calm my fears. It was humbling to learn to accept others and myself for our struggles, something I had trouble with before. I thought a person was nothing if he/she wasn't perfect." Sometimes people don't distinguish between having high standards, which give you pleasure, and expecting perfection, which will eventually lead to pain and disappointment.

Cynthia wrote, "Sometimes when I'm anxious, I forget to breathe. But, it's getting easier to remember the way we helped each other at group meetings. As soon as I remember, my breath flows and brings me to a place where I can feel my inner strength. I get calm and feel an inner peace."

"I learned to trust and let go of needing control. Trusting my doctors was important but now I'm most aware of trusting myself, which is even more important as a mother."

"I was able to make a bold choice to reject my nine-hour-a-day job after working so hard to get my daughter. I worked from home at first, but ultimately I abandoned my great-paying job and career and became a dance teacher again. Spiritually, physically, and mentally, it's been the BEST thing I've done since having my daughter."

## Gaining Clarity about Work

Cynthia's choice brings me to the work issue. Job-related lifestyle is a big factor. Over the years, I have watched women change jobs or take a leave of absence from their jobs because they become aware that it is next to impossible to be slave to two masters (a paying job and managing the unpaid full-time job of infertility treatment). Here is where lifestyle changes can make a vast difference either prior to motherhood or after it.

Some women want to work, some women need to work, and some women choose not to work. There is no right answer to the question. What matters is that you feel the inner resonance with your truth, as did Cynthia.

The issue of resonance with work resolved for Shari when she let go of her demanding job as an event planner. When you met Shari in chapter 4, you learned that she and her husband Steve resolved the fertility crisis with the adoption of their daughter. But, the opportunity to learn from adversity did not end when they brought Mai home.

In my mind/body group, Shari became aware that the intensity of the stress resulted in major stomach problems and acid reflux. She had learned to control her anxiety and her physical symptoms. In addition, she took a yoga teacher training course as a way of grounding herself in serenity. Shari wanted to maintain this serenity after the adoption, but she realized that the job was "crazy-making." She also realized that she was afraid to leave the security of it, but now she had an option.

Shari was able to draw on a variety of cognitive mind/body skills that she had learned in the group. She was able to access a more optimistic willingness to take a risk. She began to trust that she could make a living teaching others to stay centered and, as a by-product, have her own centeredness as a way of life. She was able to integrate a faith in the future. "I realized that if the yoga thing doesn't work out, something else will."

It has been exhilarating for Shari to be able to take that leap of faith and claim the double benefit of loving her work and feeling physically well. She loves having more time with her daughter. Teaching yoga is "good for my students, good for me and my family, good for the universe—so what if it is not as much money? At least I'm not paying for it with that horrible reflux. I've never felt in the driver's seat of my life like this before."

There is special reward for Shari because she offers yoga classes to others who are under the strain of waiting for an adoption, a strain that she knows all too well.

It was common for women like Kim to take time off from work *while* trying to conceive. Many found it easier to concentrate on experiencing what their bodies were "saying" to them by eliminating, taking a hiatus from, or reducing the stress of work.

Arriving at parenthood inspired some women like Julie to concentrate on being a parent. She vowed that she would return to work some day, but *only* in a job that matches her authentic value system and resonates with who she really is. It makes a big difference when you take more out of your job than it takes out of you.

Also important to Julie, was that motherhood allows her to continue healing her femininity, which was marred by her anorexia. Unbelievably, Julie was actually grateful for her sixteen-week miscarriage because the emotional growth that ensued helped her get over the fear of living in a female body. For her, the pressure of a job outside of the home would have squeezed out the space for this important accomplishment.

Julie shares the decision to be a stay-at-home mom with several of the women whom I interviewed. Lauren, an experienced prosecutor, realized that her "heart (key word) is no longer in this job." Like Julie, she plans to hang out with her son for a while and when she goes back, it will be to a job that she says "[her] heart will lead her to, based on [her] changing sense of purpose. She said, "I feel like I will be led into the nonprofit sector where I can help underprivileged children or battered women—some kind of work that helps people in crisis."

Cecile, another ex-lawyer, quit the legal profession at the outset of the fertility struggle to follow her passion to become a writer. For her, the legal field would have made it more difficult to keep balance in her life. She also feels pleased to be able to give back by sharing her story in this book.

## Benefits Big and Small

Gratitude was at the top of the list of what virtually everyone who participated in this book said they gained. In one way or another, each woman spoke of an awareness of how they take the trials and tribulations of being a parent in stride now because of the arduous journey to their children. With an easy conception, some felt that might not have been the case. The level of patience they needed in order to deal with the infertility had increased and was serving my interviewees very well as parents.

Sarra learned soon after the birth of her daughter, how much the template for dealing with adversity, which she developed before conception, prepared her to cope with the sudden death of her father. Like Evelyn and Robin, Sarra was no longer stuffing her feelings. Lauren, Robin, and others learned that they couldn't expect their families of origin to be different than they already are. What matters is that they can live their lives as *they* choose, despite the opinions and attitudes of others. Many became aware that an increase in self-esteem ranks up there on the list of gains.

Shari mentioned that it feels good to have learned how to be kinder to herself. "Never would I have allowed myself to 'laze around' reading—a yoga book, no less!—before discovering the power of inner stillness." When the mind and the body are spewing static electricity everywhere, and when uncertainty rules the day, coming in for a landing in our inner sanctuary, in the present moment, quells the frenzy and allows for self-awareness to take us to our truth.

All participants experienced a deepening of new and old friendships and relationships. They clarified priorities about work and parenting as the passion of the heart found its free expression. Awareness of troublesome ego states allowed for what Brenda called "rewriting history." Many became aware of their capacities to screen out negativity and create joy. They embraced humor as an effective coping mechanism. Stacie, for instance, joked about the fourteen embryos magnetized to their refrigerator, and how they were planning to put fourteen stockings on the mantle at Christmas.

## Shit Makes the Flowers Grow

Infertility is a shitty experience. Can you embrace this metaphor and allow this shit to fertilize your bloom? At the outset of the experience, you may not be in the mood to care if there is anything to gain. I understand. Now that you have read this far, I hope you can appreciate that being the evolver *and* the evolvee keeps you busy in a productive way while you put one foot in front of the other through this endurance test.

## Be Warned: Growth Is Not a Straight Line

There are two things to keep in mind about the growth process. One is that it is most common for the gains that we make to be vulnerable to the gravitational force of what had been habitual. Therefore it would be a mistake to interpret a lapse as a relapse.

Let us say that you have arrived at a place where you are pleased with yourself for a new and consistent capacity to take the stupid comments that people make more lightly. Suddenly, you feel mired in a reckless remark that you cannot seem to let go of, for instance, "Get drunk and you'll get pregnant."

This could be good news. No, I have not lost my mind. The growth process goes back and forth. In fact, it is victory that provokes a U-turn back to the habitual. Remember that an old ego state has become "unemployed." It pulls back and regroups, waiting for a vulnerable moment to reassert itself. If you *believe* that the slip-up is a sign that you haven't changed, then that old part of you gets its job back. It is important to remember that feelings are not facts. What part of you is having the feeling? Use the hard-earned, changed part of you to take a firm stand and reclaim your victory. People who make stupid remarks do not deserve the energy it takes to thrash around with them.

The other important detail to keep in mind is that there will always be leftovers—unfinished business or new goals—that present themselves as your self-awareness continues to grow. All of the women who had gained so much were able to list what still needed attention. A few mentioned the need to repair the damage to their sex lives, for instance. Others mentioned personality traits that have become conscious that they want to understand and rework.

Some identified traits that had improved but not enough to satisfy. There is always room for improvement. Such is life.

## Exercise

Spend some time in quiet contemplation. Identify something that has already changed. Close your eyes and allow yourself to have a daydream in which you see yourself using this new skill under varying circumstances in the present and in the future. At the end of the exercise, use the space below to capture your musings. Then make a list of what you'd like to gain next.

Chapter 11

# Give Birth to Yourself: A New, Improved You

## Become a Pro at Creating

Procreation is the most profound form of creation. It is your goal. Unfortunately, you're on an obstacle course that is also an endurance test. The rigors of treatment, or time in between treatments are best balanced by entering an infertility-free zone. Other kinds of creativity can be a real boon at this time. Whether enjoying someone else's artistry or getting lost in your own, the arts are a wonderful place to dive under the turbulence and find an infertility-free zone.

Absorption in creative endeavors is a wonderful distraction and makes the time passage of time productive. Anything that smacks of creativity can keep the spotlight on joy. Joy is what is lost in this process, and the creative arts are an excellent way to shift the focus back in that direction. Creative endeavors combine the best of doing and being.

Robin, for instance, told me that she and her partner "bought a small, fixer-upper vacation house, so that the infertility wouldn't dominate." One need not have the money to buy real estate. Buying a few yards of fabric to make throw pillows or placemats can add spice and distraction to an emotionally difficult day.

Jenn, the singer/songwriter, wrote a number of songs about longing for something that you don't yet have. Her words to me were, "and I'm going to give birth to that album, too."

On a more mundane level, I can remember a time when I was agitated about something, and could not come in for a landing in a meditative place, but I knew that I needed to. I remember grabbing a sketchpad and a few crayons and mindlessly doodling. The next thing I knew, it was forty-five minutes later, and I felt much better. Doodling took me under the turbulence.

Music can serve this purpose as well. I teach an experiential lesson in my mind/body classes. While participants hold a distressing memory in their minds, I play a jazzy, upbeat piece of music. Music is one of the languages of the limbic system where healing takes place. Participants spontaneously begin to "dance" in their chairs. They learn that they can spontaneously shift to a different perspective.

I heard a fascinating story on National Public Radio's *Weekend Edition* a while back. It was about a project in which inner city kids in Los Angles were paid to come to a class on Shakespeare. Their "job" was to understand what Shakespeare was saying and to reimagine his work so it could feel relevant to their lives. Teachers guided them through Shakespeare's "foreign" language to the meaning of his plays, sonnets, and other works. The kids teamed up or worked on their own to create a song, poem, rap, or other work of art. These were collected on a CD called "Urban Shakespeare." The consensus of the participants was about how empowered they felt at what they had created. Empowerment is the perfect antidote to helplessness.

The interviewer in this story was Christine Anthony. Mary Beth Kirchner, director of education in Los Angeles, spoke with Christine about the Shakespeare project. What Christine said was memorable: "War pumps up the need to create. The opposite of war is art." Inner city kids feel as if they live in a war zone. *So do infertility patients.* My interviewees all used the same language to describe this experience: in the heat of battle, trench warfare, battleground, and the like. Art is the opposite of war.

Can you locate your inner artist and give her/him expression as an antidote to this temporary war? Can you apply that creativity to the built-in challenges of infertility?

## When Is Enough Enough?

Whether the issue is shifting to another option or stopping the quest altogether, you want any decision you make to feel real and authentic. Jenn observed this pitfall and became curious about those who go through "seven, eight, or nine IVFs." She marveled at the stamina needed to go that far, but based on her experience with IVF, she theorized about how she thought these women could persevere.

She said, "You start one cycle, and it is overwhelming. You don't know what to expect. And, the second time it's like 'okay, this could work this time.' So you start taking your shots. There's a tad bit of excitement in it and hope, and then there are all of these hurdles that you have to get over. Oops, got over my estrogen hurdle—my levels are good. Okay, now we go for the first ultrasound to see how many ovum are maturing. Oops, got over that hurdle. Oh, my body's doing good this time. So you have this thing almost like you're in a race, and you're jumping and going. You gear [up] for the next IVF and the next, but there's like this little addiction going on. Okay, here I go again. I don't have to, but I'm choosing to. There's part of me that has to. I want to because I could actually get to the 'high.'"

Do you think Jenn's theory has merit? If this sounds like you, just make sure that you're continuing to pursue pregnancy for reasons that feels right to you consciously as opposed to being on autopilot.

Some couples can wrap their brains around alternative ways of getting a family easier than others. Joyce and Bruce needed to go to the very end of what modern medicine had to offer before they could shift to the sperm donation option. Resolution of a severe male factor diagnosis took them through the world of medical science fiction. Enough was not enough until this couple was sure that they would not to look back in regret. They can now be fully content with their two children conceived with donor sperm.

## Child Free Living

How will you know when it's time to stop altogether? In the introduction to *On Fertile Ground* I promised you more on child free living. I suggested that if you read this book you would have the tools to shore up your stamina. But, at some point you may feel that you've dead-ended. What happens then?

Sue Slotnick and her husband, Irwin, arrived at the child-free decision but, of course, not easily. An important player and colleague of mine in the national organization, Resolve™, Sue recommends living for a while as if child free *is* your choice. This will help you to clarify if you can truly live with the decision.

It is important to differentiate ending your quest because it's right for you, as opposed to giving up. Dawn Gannon, another Resolve colleague, eventually settled on the child free choice. She points out that you and your partner both need to be on board, and you may need to allow time for you to catch up to one another as was the case with her and her husband. These people wanted children as much as you do, but ultimately embraced child free living as the best choice for them. And both of these couples have children in their lives whom they adore.

A more difficult situation ensues when you involuntarily find yourself in the child-free category for one of many possible reasons. Whether by choice or involuntarily, at the very least, you owe yourself some counseling, which Sue and Dawn tout as being an invaluable part of coming to terms with the loss of your dream.

Pathos is part of life. The potential for loss hovers over all of us. No one likes this fact. The question is, "How do we make peace with an unasked for reality?" How do we find what Loretta La Roche (the comedienne who has been a part of Harvard's Mind/Body program) has called "the bless in the mess" thereby staying open to the joy in life? Not easily. There will always be a place where what Sue calls the "sloppy wet kiss factor" will be missing. She recommends looking at the advantages. If this is your fate, how can you use the advantages to live creatively?

Child free living makes room for the important ways in which creativity of other kinds can be so fulfilling. The options are limitless. Can you lead yourself through life by the excitement of your creative juices? Your capacity to metabolize grief and invest in joy is the path to living without bitterness.

## Advanced Maternal Age

Creativity figures into the age dilemma too. It is particularly upsetting to be told, when you feel in the prime of your life, that you are too old to have a baby. As women, sore breasts and flowing blood, as predictable as the phases of the moon, are constant reminders that we are built to reproduce. I can remember many years ago reading an advertisement in the *New York Times* that one of the clinics was looking for ovum donors between the ages of twenty-one and thirty-five. I was thirty-seven at the time, and I remember feeling horrified that, had this been my inclination, I would have been too old. These days the late thirties and early forties are certainly years in which you are youthful enough to chase after little children. There was a time when the thirties and forties were the life expectancy. Humankind has evolved from a way of life that is now obsolete.

Many years ago, I read a passage in James Michener's book *The Source*[50] that made a deep impression on me. Set in Israel at an archeological dig, Michener launched into a description of an olive tree that was many hundreds of years old, and mostly necrotic. Yet, there was a proud, tall branch, reaching for the sun, which was still bearing fruit. Who is to say that this metaphor will not be your reality?

Sarra, who interestingly has the name of the woman in the Old Testament who gave birth later in life, trusted her intuition that "the statistics might not be interpreted properly." She went on to not only have her beloved daughter, but, as she put it, "to have a pregnancy like a twenty-year-old. I swam for exercise until one week before I went into labor. I was enjoying every moment of my pregnancy, and I had a normal delivery."

Some medical options hold that Sarra's story contains an irony. If she had not searched and found the medical team who diagnosed her immune situation, and if instead she had abandoned the quest for a genetic baby and had moved ovum donation into the number one slot, her level of killer cells were so high that ostensibly any pregnancy achieved by this method might have been unlikely to survive. Maybe, in the final analysis, both Sara from the Bible and Sarra from New York were just plain lucky. Some might say there was divine intervention. There are things we just don't know.

## Ovum/Sperm Donation

Ovum donation is an amalgam of creative realities. Just the idea that this is a choice is astounding. Those who donate their "seeds," even if it is for money, still get credit for their participation in a creative act. To claim this choice as your own takes creative participation, under the turbulence. There is a new field of science you should know about called epigenetics. It has shown that the development of an ovum-donor baby in your body will be influenced by you physiology even though the baby does not have your genes. Some candidates find this news makes it possible to accept this choice.

As you must know from the stories in the news, the likelihood is that you are not too old for ovum donation. I can only encourage you to trust this creative choice after you've soul-searched to determine if it's right for you. A growing body of literature and resources can support you in this choice. A good place to start is www.resolve.org.

On a personal note, I am proud to say that three of my four treasured granddaughters are ovum donor babies. Chemotherapy and radiation destroyed my older daughter's fertility when she was seventeen. Jaden, and twins Macy and Reese, will always be a source of wonderment and delight for which I shall be eternally grateful.

Sperm donation has been around for a lot longer than ovum donation. But any choice, other than the old fashioned way of conceiving, takes an investment in creative thinking to come to the decision and to trust that you can learn the best way to deal with your child's questions about his/her origin.

## Lesbian/Gay and Single Parenting

I can sum up the essence of these creative choices by quoting James Baldwin. In his book *Giovanni's Room,* he says, "If it's love, God won't mind."[51] Good for you if you have the courage to be in the minority and openly love someone of the same sex. And good for your child, who will be lucky to have your love.

## Adoption

One of the greatest calls to creativity is adoption. I have been greatly moved emotionally by families in my practice whose choice was adoption. My first experience was indelible. I felt privileged to be privy to Katie and John's joy as they brought their little Lindsay to meet me. They said, "Helen, look at her—we could not have done better ourselves."

Shari and Steve are completely smitten by how "fulfilling and rewarding" their adoption experience has been. Shari told me that she wished she had realized this sooner during her ordeal. She would have chosen adoption earlier. Friends who observed them with their daughter Mai have said, "You'll never win the lottery because you've won it already."

Jenn came to terms with becoming a mommy by adoption when she really understood that she wanted a family more than she wanted a pregnancy. Then she found herself in an unlikely and surprise pregnancy. She claimed that if she does not conceive quickly with baby number two, she and Josh would look forward to welcoming an adopted baby into their home.

There are aspects of infertility that have no rhyme or reason. Jenn and Josh, who had abysmal odds, did conceive again—on their own, no less—and their son now has a sister. The outcome of the fertility quest can be as predictable as a crapshoot. Who knows why the dice were loaded in their favor? I hope that they're loaded in yours.

Beauty and creativity can be completely congruent. Nanci and Joe have two adopted sons from Colombia. Their story was orchestrated with creativity and oozes beauty. Nanci told me that when they decided to adopt, both of them were clear that they could "teach a child to love being from Colombia." They are committed to keeping the cultural inheritance of their sons in the forefront. They have prepared "life books" for each of the children, books that "tell the story of how they came into our family. [They] review [them] with [the children] frequently."

Nanci told me, "We find ourselves now as sort of ambassadors for Colombia. We try to change peoples' opinion of it because we are very aware that our kids are going to be growing up in a culture (the Deep South) where people have a negative opinion of the country that our boys come from. We work hard to re-educate people whenever we hear negative stereotypes. We watch the news of Colombia very closely. We cook food and have art from Colombia in our home. The kids will have Spanish lessons as will Joe. (Nanci speaks Spanish.) We intend, as the boys get older, to make regular trips to Colombia to keep them very familiar with their heritage. And we've chosen a place to live based on a school district that provides as much diversity as can be found in this neck of the woods." Nanci said, "We are doing it for them, but it's an amazing gift for us, too, a gift that has enriched our lives immeasurably."

## From Road blocks to Building Blocks

The premise of this book holds that the road blocks to fertility can become building blocks.[52] The building blocks, mind/body coping skills, allow you to construct a new version of you. Living with greater satisfaction, greater confidence, greater optimism and greater clarity are the gains from the pain.

Taking the steps necessary to arrive at parenthood has become an organizing force in your life. Time may feel like it is dragging on, but you have a chance to spend it purposefully and productively as well as creatively. Until you bring your baby home, practicing self-care coping is about conceiving of and nurturing your higher potential. No one can decide for you if this is a valuable pursuit. If you have read this far, I presume that you are at least entertaining the notion.

Growing from the adversity of infertility has not been central in the literature. When Monique was in her five-year fertility quest, she read every book she could find on the subject. She said that she found "a gap between books that present the physiology of infertility and the things that you can do to conceive (a western mindset) and books that present spirituality, logic and a philosophical orientation focused on making meaning out of suffering (a typically eastern approach)." She went on to say that "these eastern books are good jumping off points but they don't get to what dealing with infertility can do for you if you assume responsibility and are *disciplined about it*."

Bingo! "Disciplined about it" is exactly the point. Discipline sounds like a chore, and I understand how unappealing that might seem when you are feeling oppressed by what you are going through. As Monique and I spoke, she told me why she jumped at the chance to be a part of my project. From my e-mail invitation to participate, she said she could tell that "this book would provide the bridge" between what she saw as an East/West polarity. Yup. That was my intention.

Monique told me the books that frustrated her the most "ended with 'and then you have a kid, adopt, or neither' but *they do not consider the ways in which infertility changes your life.*" She said, "It's the discipline that allows you to discover that you can sit down to meditate everyday for three months and nothing will happen and all of a sudden you have a huge insight." Insights are one ingredient in the recipe for turning the tables on change. Embracing mind/body options provide relief and teach you about yourself.

## Locating an Infertility-Free Zone

Every minute can feel like an hour when you are suffering. Relief can come from balancing the intensity of infertility with the pleasures of creativity. And it can come from the discipline to dive under the turbulence to an internal oasis, thereby breaking the spasm of stress. Your body returns to neutral, even if temporarily. It's easier to "pull up your socks and get on with it" if you can do something that returns you to neutral so that you want to pull up your socks.

In your situation, I believe that creating a mind/body state that is as close to neutral as possible is an important concept. Spending time following your breath or your thoughts or a word, phrase, or prayer, or engaging in mindful walking or mindful cooking, for that matter, all have the potential to bring you back to yourself. All of these techniques are meditative. All shift you in the direction of neutral.

*When it comes to meditation, I believe that a loose definition and low standards are good enough.* People sometimes give up because they feel like they are failing if they cannot sit as still as a Tibetan monk after struggling to focus for three or four minutes.

Anything that breaks into the frenzy of infertility is an important victory. The study on letting-go coping that I cited on the very first page of chapter I suggests that relieving mind/body stress and landing in a place of relative tranquility could play a part in creating receptivity for a conception. Stacie was thrilled to discover that she neutralized the feeling of being overwhelmed the time that she disciplined herself to do yoga for forty-five minutes. She had to get past her inertia to get there, but it provided an "aha" moment that was well worth the effort.

I received a call recently from someone who appeared to be interested in my mind/body group. "What was it about?" she asked. When I told her she said, "I know all about that inner-stillness stuff. I never got there." And she hung up. Here was a woman who must have heard a rumor that diving under the turbulence could help, but did not understand that she needed more guidance, persistence—and, yes, discipline—to get to that internal oasis that for her must somehow have become a strange, taboo destination. She never gave herself a chance to have the experience that Monique had: "that one second of serenity and then [she] was hooked."

Reproductive technology is awesome. *But so is the empowerment that we feel when we enter the place below the turbulence of infertility where our minds and bodies can inform each other.* Several of the women featured in this book had received messages intuitively from under the turbulence and knew that they could conceive and carry a pregnancy on their own. Others came to terms with adoption or accepted the option of donor sperm or eggs from their internal oasis. Our wisdom awaits us in our inner infinity. Important insights and answers can flow into the void we create when we disengage from worry.

In today's world, it is easy to feel removed from connection to ourselves. We expect ourselves to be master multitaskers. Anything that slows the warp speed of our lives gives us a chance to transform adversity. But we must have, as Monique says, the discipline to experience our inborn capacities for internal focus. Otherwise the external world, which is like a noisy carnival, will seduce us into its chaos. In the process, we can give birth to ourselves.

## Infertility Causes Stress for Sure. But Does Stress Cause Infertility?

It is a no-brainer, and universal, that stress levels elevate because of infertility. The diagnosis and its aftermath create a fear response that tightens the body. Letting-go techniques loosen the tightness and as such can be a low-tech solution for some couples.

My take on this ongoing debate has to do with making an important distinction between the stress of infertility the stress of traumas that may be lodged in our bodies. A new study claims that levels of cortisol, a stress hormone, are higher in women who are not conceiving.[53] No kidding. We need to leave room to take unconscious factors that can be healed into consideration.

Stress can be related to what's going on in the outside world, but it's important to understand that stress is not an absolute but a matter of conditioning and perception, in which case stress comes from the inside. Does stress cause infertility is not an easy question to study.

While there are factors that it might not be possible to know, I would state my opinion this way: *The stress of not being fully in touch with the truth, not being in touch with unconscious processes, might block fertility.* The stories below leave room to imagine that emotional blocks can be released when an *underlying belief system is acknowledged.* Nature loves the truth. If you plant a tulip bulb upside down, the shoots will make a u-turn so they can reach through the soil to the sky. Getting to your truth can happen with the discipline and dedication to crossing the bridge from what to do, to how to be with yourself. Your wisdom and intuition can be freed to reveal themselves. Worst-case scenario, you may need to seek professional help to unearth what is lurking just below conscious awareness.

Many years ago, a woman came in to my office and before she had even sat on my couch, she asked me if her failure to conceive could be her fault. I said that I was not fond of people blaming themselves, but I was curious as to why she thought this. Out tumbled a story about a childhood urinary problem that was finally resolved surgically when she was sixteen. Her mother had been harsh and critical, irritated that her daughter wet the bed nightly.

I asked her if she was afraid that she would *be* the kind of mother that she *had.* She burst out crying. Shortly thereafter, she conceived. There is room to imagine that worrying that she was doomed to be just like her mother was stuck in her cells. Apparently, her tears and other aspects of our session released that worry and freed her body enough from this concern for her to conceive. I assured her, based on my years of experience, that those who stake a claim to depart from the way they were raised, and then follow through (with support, if necessary) have a great shot at not repeating history.

Brenda is a living example of this. When I asked her why she thought she got pregnant twice on her own after her first baby was conceived with IVF, she said, "I think maybe my first pregnancy opened up what was not working and then the emotional block was gone." In my opinion, once Brenda realized that reaching out for child-raising guidance as soon as she gave birth to her first daughter relieved the concern that *she* would be the kind of mother that *she* had.

And remember Lauren? When I suggested that she might not want a child because she already had a very difficult "child" in her demanding mother, she was greatly relieved to hear me say that because she had that thought but dismissed it many times. We exposed the underlying belief to the air, and, like anaerobic bacteria, it died. She conceived and carried to term shortly thereafter. Amelia and Angelo thought that it was a blessing, and maybe a factor, that they had miscarried when their relationship was so toxic.

Life is stressful and people conceive in spite of it. No one should *ever* be blamed for their fertility issues. Women have expressed regret that they didn't start to try sooner. If they had known, they would have. Women are smart enough to wait until the right man comes along. Marrying late is not your fault either.

Some problems with conception are congenital. How ridiculous to blame yourself for having blocked tubes, a T-shaped uterus, or for the fact that your mother may have taken DES.[54]

A compromised sperm count could be the result of having the mumps, but, could it be a mind/body expression of lousy fathering or mothering? A man's body is not exempt from expressing a mind/body conflict. To the men reading this book I say, you might want to follow this hunch if it resonates for you. But, that still doesn't make your infertility your fault. As with women, it's not that stress would cause infertility, but rather that unconscious **un**truths are stressful.

An unresolved abortion might need to be grieved in order to be released. I've worked with many people who are consciously clear that it would have been a disaster to have had an unplanned baby. But, on an unconscious level, another story was not letting go of its demand for deeper resolution. This is exactly why self-awareness is so important and so healing, and why letting go can take you deep enough to make important discoveries.

Whether you trust a higher power to get you to your goal, or, like Carol, your doctor's brain, it is critical to select a medical team in whom you can have confidence. Modern medicine works miracles with your "seeds." This has been a book about how you can tend to yourself, the "soil" in which those seeds must take root. It is you who are the fertile ground.

## When You Arrive at Parenthood, Does Infertility Feel Resolved?

Opinions varied on whether infertility can feel completely resolved. Here's where we need to look at the issue of trauma. Infertility is traumatic every which way to Sunday. But all of us have been traumatized by life, too.

Some claim infertility is in the past. Sarra said, "I have graduated now." Others feel that there is a residue that could never be swept away. When I asked Cecile if she felt on the other side of the fertility struggle she said, "It's diminished. In some ways I still consider myself infertile because I don't know if I can ever have another child." Then I received an update: Cecile in fact conceived without intervention and has a second son. She says, yes, her infertility now feels resolved.

We cannot "un-have" our history. Cecile's email went on to make this point. Now a writer, Cecile told me that she "interviewed a woman for a magazine pitch on infertility, and [her] feelings and experiences brought back so many bad memories that...[she] sat in her boys' bedroom for a full hour watching them sleep, [her] heart aching for the woman with whom [she] had just spoken."

Nanci's answer to the question is the infertility in the past is also yes/no. She told me that she was "absolutely" on the other side of her infertility. But later on in the interview she shared that she "did not feel one hundred percent healed" when she watched her parents making such a fuss over the pregnancy of her brother's wife. The difficulty intensified each time the brother e-mailed the sonogram pictures to them because Nanci miscarried after they had seen the first sonogram of the pregnancy that Nanci lost to miscarriage before she and Joe decided to adopt. Her brother's insensitivity was a trigger for Nanci. No matter how healed you feel, triggers are inevitable. But that doesn't mean that you are not *healed enough.*

There's more to the story for Nanci that I believe will clarify things for you. She told me that she gets disoriented when someone who has adopted goes on to conceive. She is also mystified when someone conceives easily with IVF. But, the good news, according to her, "is that I am so pleased that through my work with you I came to know myself and identify that *it is the residue of the perfectionist aspect of my personality that haunts me at these times.*"

Nanci's insight is central to understanding the role of trauma. We need to distinguish between resolving the fertility (which means you get your family or you come to terms with living child-free) and understanding that infertility is traumatic and *the residue of traumas leave their mark.* Traumas have a way of getting lodged in our bodymind. A sister-in-law's pregnancy would have had to touch a vulnerable place in Nanci. When we are vulnerable, old aspects of our ego elbow their way back into the driver's seat. So, you can see the logic in her feeling that she is "absolutely on the other side of the infertility," and then describing the episodes that leave her feeling that "she is not one hundred percent healed."

The key to her actually being as healed as anyone can be, is that Nanci recognizes that the perfectionist aspect of her personality, when provoked, leaves her feeling that since she didn't conceive and carry a biological child, she's not perfect. If you develop Nanci's level of self-awareness, then moments of vulnerability can become "aha" moments and can be used to invite the old part of your ego to go back to the viewing stand, so the new you, with an authentic point of view, can be leading the parade. Nanci doesn't let the ANT, "I'm not perfect," spoil her picnic.

Remember Kelsey who never thought she'd say, "If I had the choice, I wouldn't change a thing"? In her case, a simple reframe from her husband shifted her bitterness and healed that wound. But, since the limbic system holds traumas, the determination to work through your trauma will get you to the other side. If you find that you cannot get to this kind of insight on your own, it would be wise to reach out for help.

## Back to the Issue of Change

*"Relax, honey—change is good."*

When all is said and done, the universe may have placed roadblocks in front of you that seem insurmountable. Something, and maybe many things, must change for you to fight this battle without getting maimed. Embrace change. It may feel dangerous, because you need to go where you've never been before. But keep in mind that it's also an opportunity, pregnant with the possible.

The choices are many for resolving this crisis. With all my heart, I wish you good luck in making your way to your true self, good luck in making your way to your children and for some of you, good luck in making your way to living a creative, child free life. Armed with mind/body coping skills for navigating life's trials and tribulations, no one is a failure.

If the opposite of war is art, the opposite of the fear that comes with the diagnosis of infertility is love. An open heart can radiate light on, and receive light from, the blessings buried within adversity.

# Resources

These days, most people can maneuver around the Internet and find what they need. Any of the major clinics in your area will have a Web site filled with *medical* information. You may find satisfaction to unanswered questions if the site has a *search* function. Dr. Frederick Licciardi has a very comprehensive blog that maps out the various diagnoses and treatment approaches (visit www.infertilityblog.blogspot.com).

My concern is that you are able to find guidance for locating support for the *emotional* component of infertility. There are many articles in the Baby Manifest-O™ blog tab of my Web site, www.mind-body-unity.com. You will at least get an overview of some of the issues that pop up. I also offer a free twenty-minute phone consult at 212-758-0125.

When searching the Internet for emotional support, I recommend that you put the following key words in a search engine such as Google:

Emotional support for infertility
Psychotherapy for infertility
Mind/body stress reduction for infertility
Couples counseling for infertility
Hypnosis for infertility

You should follow all of these entries with a dash and then enter your geographic location (for instance, emotional support for infertility – New York City.

If it's hypnosis you want, you can go to www.ASCH.org. You can search for a practitioner in your area by designating the state in which you live.

You will have many options by visiting Web sites, such as www.Resolve.org, www.theafa.org, or www.inciid.org. These not-for-profit organizations offer a wide array of information, tips for coping, blogs, webinars, teleconferencing, support groups, and advocacy.

A very popular blog for general infertility is www.stirrup-queens.com. There are sites for adoption, surrogacy, single motherhood, child-free living, gay and lesbian parenting, and what is called third-party parenting, which involves the use of sperm, egg and embryo donation (in the event both partners are infertile). In the New York City area, there is a monthly meeting for those who have children by these methods or who are considering doing so. This name of the group is the Third Party Parenting Network (TPPN). For information about this group, contact Nancy Kaufman, MSW at 212-722-1200 or Elizabeth Silk, MSW at 212-873-6435.

Finding the right resource depends on whether you feel that you need face-to-face contact or prefer to reach out from the privacy of your own home.

Remember, information is a wonderful problem-solving, coping method that will contribute to a reduction in anxiety.

NOTES

# (Endnotes)

1 A D & C (a dilation and curettage) is a minor surgical procedure in which the inside of the uterus is scraped clean.

2 The ritual, which I created, rested on scientific wisdom that "rites of passage are psychologically…necessary to help us move forward and invest our lives with meaning." From Abigail Brenner, *Transitions: How Women Embrace Change and Celebrate Life* (Charleston, South Carolina: CreateSpace, 2010), 5.

3 Ellen Frankfort, *Vaginal Politics* (New York: Bantam Books, 1973).

4 Nathalie Rappoport-Hubschman, et al., "Letting Go: Coping is Associated with Successful IVF Treatment Outcome," *Fertility and Sterility* 92, 4 (2009): 1384–1388.

5 Rappoport-Hubschman, 1386.

6 A statistically significant statement is one that has a greater likelihood to have a true causal relationship than what might be expected to happen by chance alone.

7 Eliahu Levitas, et al., "Impact of Hypnosis During Embryo Transfer on the Outcome of In Vitro Fertilization– Embryo Transfer: A Case-Control Study," *Fertility and Sterility* 85, 5 (2006): 1404–1408.

8 The bodymind is a term first proposed by Dianne Connelly.

9 Tara Parker-Pope, "Learning While You Dream," *New York Times*, April 22, 2010, Science Section.

10 A.D. Domar, PC Zuttermeister, R. Friedman, "The Psychological Impact of Infertility: A Comparison with Patients with Other Medical Conditions," *Journal of Psychosomatic Obstetrics & Gynaecology* 14, supp (1993): 45–52.

11 Elizabeth Kubler-Ross, *On Death and Dying* (New York: Scribner & Co., 1969).

12 Follistim is the trade name for FSH (follicle stimulating hormone), which is given by injection to stimulate the ovaries to develop multiple follicles. It is made in the laboratory, and it can be used an IUI or an IVF cycle. Follistim is manufactured by Ferring, and a very similar FSH product, Gonal F, is made by Serono.

13 hCG is Human Chorionic Gonadotropin. This is the hormone that is made by the early pregnancy and then the placenta. A urine or blood pregnancy test measures hCG. hCG has another important role in fertility treatment in that we give it by injection to start ovulation. In the natural cycle, it is the LH (leutinizing hormone) that is released from the pituitary gland to initiate ovulation at mid cycle. The two molecules, hCG and LH are very similar in structure, therefore hCG can be given by injection to mimic the LH surge. hCG is used instead of LH because hCG is much easier and less expensive to produce. hCG can be used to trigger ovulation in a natural, clomid, or injection cycle.

14 Sharon Begley, *Train Your Mind: Change Your Brain* (New York: Ballantine Books, 2008), 9–15.

15 FSH is the abbreviation for follicular stimulating hormone, which stimulates the maturation of ovum.

16 Stressed is desserts spelled backward is a phrase coined by Loretta La Roche, a comedian who teaches in the mind/body program at Harvard's Benson-Henry Institute for Mind/Body Medicine, Boston, Massachusetts.

17 Harville Hendrix, *Getting the Love You Want* (New York: Henry Holt & Co., 1988).

18 DQ alpha is a protein complex found on the surface of white blood cells. It is one of the proteins necessary for the normal function of our immune system.

19 Antiphospholipids are antibodies that can be found in many people with autoimmune disorders such as systemic lupus, rheumatoid arthritis, Sjogran's syndrome, and others. Phospholipids are the necessary fat molecules that make up the membrane of cells. Some people with autoimmune diseases make antibodies against the phospholipids in their own cell membranes, thus the antibodies destroy their own cells.

20 *Finding the Eye and the "I" in the Storm* is the title of an article I wrote that *Family Building Magazine* published in the fall of 2003. The title is not only about the need for respite, but also because the I gets shattered in the fallout of the infertility struggle.

21 FSH (follicular stimulating hormone) is a hormone, made in the pituitary gland, that stimulates the follicle to grow each month. The follicle is the fluid-filled structure that holds the egg; each egg is in a follicle. If there is a good quantity of eggs in each ovary, a woman's FSH level on day two or three of her cycle will be in the normal range, less than ten, the lower

the better. When the number of eggs in the ovaries diminishes, FSH levels increase. The body can sense that the egg reserve is lower, so it tries to drive the ovaries to do better by increasing the amount of circulating FSH. In menopause, when there are no more eggs available, FSH levels can be as high as 100.

22 Leukocyte Immunology Transfer is a controversial, inconclusively tested approach to protect a fetus from immune rejection when a woman's inability either to conceive or hold a pregnancy to term is related to the functioning of her immune system.

23 A biochemical pregnancy is a pregnancy that stops growing a short time after conception. There is a positive pregnancy test, but the pregnancy never develops enough to become viable and seen on ultrasound. Within a few days to two weeks of the first test, the hCG levels fall away to zero.

24 Daniel Amen, MD, *Change Your Brain: Change Your Life* (New York: Three Rivers Press, 1999).

25 A hystosalpingogram is a medical procedure that determines if the fallopian tubes are blocked.

26 Endometriosis is a condition in which cells that belong in the lining of a woman's uterus appear and flourish in areas outside the uterine cavity such as on the bowel or ovaries. This can cause pain, irregular menstrual bleeding, and infertility for some women.

27 Intracytoplasmic sperm injection (ICSI – pronounced "icksy") is an in vitro fertilization procedure in which a single sperm is injected directly into an egg in the laboratory. It is used to assist the sperm in fertilizing the egg and might increase the likelihood of fertilization if fertility problems are a result of problems with sperm.

28 Sperm DNA integrity assay is a test that goes beyond determining sperm count, morphology, and motility. It analyzes sperm integrity to rule out sperm fragmentation, which can influence rates of fertilization of eggs and also subsequent embryo development.

29 Amniocentesis is a test during pregnancy that removes a small amount of fluid from the sac around the baby to look for birth defects and chromosome problems.

30 Comparative genomic hybridization is a screening technique that permits the detection of changes in chromosomal numbers.

31 Amenorrhea is where a young woman ceases to have menstrual periods.

32 Bruce Lipton, *The Biology of Belief* (New York: Hayhouse, Inc., 2005), 5.

33 Lipton, 7.

34 Lipton, xvi.

35 Candace Pert, *Molecules of Emotion* (New York: Scribner, 1997), 188.

36 Herbert Benson, *Timeless Healing* (New York: Scribner, 1996), 35–37.

37 Carol Ginandes and Daniel Rosenthal, "Using Hypnosis to Accelerate the Healing of Bone Fractures: A Randomized Controlled Pilot Study, *Alternative Therapies* 5, 2 (1999): 67–75.

38 Herbert Benson, op.cit., 39.

39 Superovulation is using fertility drugs, either orally or by injection, to stimulate the ovaries to produce more eggs than would normally be made in a natural cycle.

40 Vericoelectomy. Like all organs of the body, the testicle has arteries that bring it blood and veins that carry blood away. Veins of the testicle that are enlarged and swollen are called varicoeles. These large veins are very similar to varicose veins of the legs. Theoretically, these large veins may impair fertility by increasing the temperature of the testicle. Another theory is that the large veins restrict the ability of the testicle to quickly clear away substances that may be toxic to sperm production. Some physicians feel that removing these veins will increase sperm counts and function.

41 Sharon Salzberg, *Faith* (New York: Riverhead Books, 2002), 67.

42 Elizabeth Gilbert, *Eat, Pray, Love* (New York: Viking Press, 2006), 131–132.

43 Hyperstimulation is what happens to all women who take fertility drugs, however only a small percentage develops hyperstimulation syndrome. The ovaries are normally the size of walnuts and the fertility drugs may make them the size of lemons. In most cases, this is normal and fine because the increase in size is due to the desired increased number of maturing follicles. In mild to moderate forms of the syndrome, the ovaries become even larger, and go on to leak fluid into the pelvis and abdomen. Here women may feel bloated and crampy, their clothes may feel tight and they may temporarily gain a few pounds. In more severe forms of the syndrome, the ovaries may enlarge further, and significant amounts of fluid may leak from the ovaries, necessitating drainage of the fluid, bed reset, and, rarely, even hospitalization. In most cases, hyperstimulation can be prevented by selecting the right dose of medication and by cancelling a cycle early if the ultrasound and blood tests show that a woman is at risk.

44 Sharon Begley, op. cit.

45 Jeffrey Schwartz, *Brain Lock* (New York: Regan Books, 1996).

46 The three scales were: 1) The SCL90-R (Symptoms Check List 90 – Revised) measures perceived bodily dysfunction; 2) The LOT-R (Life Orientation Test – Revised) measures self-esteem, feelings of competency, and outlook as one score, and 3) the POMS (Profile of Mood States) measures six dimensions of emotional distress. Tension, depression, fatigue, and confusion trended downward; a statistically significant reduction in levels of anger was revealed; and the sixth category, vigor, stayed the same perhaps because of the drain of energy due to the relentless challenges.

47 Wally Lamb, *This Much I Know Is True* (New York: Regan Books, 1993), 579.

48 Giacomo Rizzolatti, et al., "Premotor Cortex and the Recognition of Motor Actions," *Cogn. Brain Res* 3 (1996): 593 – 609.

49 Clomid is also known by the trade name Serophene or its generic name, clomiphene citrate. Clomid is a fertility drug that is taken as a pill. It is an antiestrogen, meaning it blocks estrogen from getting to the very inside of cells. When Clomid is taken, the brain thinks there is no more estrogen available; therefore, the brain sends signals to the pituitary gland to make more estrogen. The pituitary gland then increases FSH production, which gets taken by the blood to the ovaries to stimulate follicle development. Clomid is a wonder drug for women who do not ovulate, as ovulation occurs in about 80 percent of such cases, which results in very high pregnancy rates. In women who ovulate naturally and regularly, Clomid is less effective, but it still can improve the pregnancy rates over baseline.

50 James Michener, *The Source* (Random House, 1965).

51 James Baldwin, *Giovanni's Room* (Penguin Books, 1993).

52 This expression was coined by my colleague, Susan Dowell, MSW.

53 Roni Caryl Rabin, "Old Maxim of Fertility and Stress is Reversed," New York Times, Aug. 17, 2010.

54 DES is a drug that was prescribed to some women in the 1960s and 1970s to reduce risk of miscarriage. DES has since been associated with increased risk of congenital uterine malformations in women whose mothers took the drug while pregnant.